# 2nd International Conference on Materials Engineering and Nano Sciences (ICMENS 2018)



IOP Conference Series: Materials Science and Engineering
Volume 378

Singapore
13 – 15 January 2018

ISBN: 978-1-5108-6585-3
ISSN: 1757-8981

**Printed from e-media with permission by:**

Curran Associates, Inc.
57 Morehouse Lane
Red Hook, NY 12571

**Some format issues inherent in the e-media version may also appear in this print version.**



Printed by Curran Associates, Inc. (2018)

For permission requests, please contact the Institute of Physics
at the address below.

Institute of Physics
Dirac House, Temple Back
Bristol BS1 6BE UK

Phone: 44 1 17 929 7481
Fax: 44 1 17 920 0979

techtracking@iop.org

**Additional copies of this publication are available from:**

Curran Associates, Inc.
57 Morehouse Lane
Red Hook, NY 12571 USA
Phone: 845-758-0400
Fax: 845-758-2633
Email: curran@proceedings.com
Web: www.proceedings.com

# Table of contents

## Volume 378

**2nd International Conference on Materials Engineering and Nano Sciences (ICMENS 2018)**

**11–13 January 2018, Hong Kong**

**Accepted papers received: 15 June 2018**
**Published online: 3 July 2018**

## Alloys and Coatings



## Mechanical and Manufacturing Engineering

IOP Conf. Series: Materials Science and Engineering **378** (2018) 011001 doi:10.1088/1757-899X/378/1/011001

# Preface

2018 the 2nd International Conference on Materials Engineering and Nano Sciences (ICMENS 2018) has been held in The Hong Kong Polytechnic University, during January 11-13, 2018. The purpose of ICMENS 2018 is promoting the creativity in the scope of materials engineering and Nano sciences.

ICMENS 2018 has attracted experts and scholars, a group of the authors and other related people attended the conference with apparent interest. We appreciate those who responded to our proposal and submitted their papers, especially those whose papers have been selected for the conference ICMENS 2018, the sponsors who have provided their valuable and professional suggestions and instructions and the scholars and professors who have spent their efforts as peer reviewers.

This proceedings of ICMENS 2018 presents papers submitted to the conference from universities, research institutes and industries. The papers selected depended on their quality and their relevancy to the conference. The volume tends to present to the readers the recent advances in the field of nanomaterials and material physics, material mechanics and solid mechanics, alloys and coatings, mechanical and manufacturing engineering, and etc. Hopeful this proceedings would be the beneficial to all participants of the conference and other readers.

ICMENS 2018 Organizing Committee
January 15, 2018

IOP Conf. Series: Materials Science and Engineering **378** (2018) 011002 doi:10.1088/1757-899X/378/1/011002

# Conference Committees

## International Advisory Committee

Prof. Richard Haverkamp, Massey University, New Zealand

## Conference Chairs

Prof. Reza Abbaschian, University of California, USA

Prof. C. W. Lim, City University of Hong Kong, Hong Kong

## Program Chairs

Prof. Moo Hwan Cho, Yeungnam University, Korea

Prof. Noriyuki Hatakenaka, Hiroshima University, Japan

Dr. Li Xia, Qingdao University of Science and Technology, China

## Local Chair

Prof. Jimmy C. M. YU, The Chinese University of Hong Kong, Hong Kong

## Technical Committees

Prof. Peerasak Paoprasert, Thammasat University Pathumthani, Thailand

Prof. Hong Yang, University of Electronic Science and Technology of China, China

Prof. Young Gyu Jeong, Chungnam National University, South Korea

Prof. Jingxian Zhang, Shanghai Institute of Ceramics, Chinese Academy of Sciences, China

Dr. Minho Yoon, Yonsei University, South Korea

Prof. Chang-Seop Lee, Keimyung University, South Korea

Prof. Ricardo Branco, University of Coimbra, Portugal

Dr. Zhauyt Algazy, Kazakh National Research Technical University, Kazakhstan

Dr. Trong-Phuoc Huynh, Can Tho University, Vietnam

Dr. Mirosław Kwiatkowski, AGH University of Science and Technology, Poland

Prof. Francesca Scalisi, University of Palermo, Italy

Prof. Josip Brnic, University of Rijeka, Croatia

Akindoyo John Olabode, Universiti Malaysia Pahang, Malaysia

Dr. Abbas Mohammed Selman, Universiti Sains Malaysia (USM), Malaysia

Chye Lih Tan, Universiti Malaysia Perlis, Malaysia

Prof. Somanathan Thirunavukarasu, Vels University, India

IOP Conf. Series: Materials Science and Engineering **378** (2018) 011002 doi:10.1088/1757-899X/378/1/011002

IOP Conf. Series: Materials Science and Engineering **378** (2018) 011003 doi:10.1088/1757-899X/378/1/011003

# Peer review statement

All papers published in this volume of *IOP Conference Series: Materials Science and Engineering* have been peer reviewed through processes administered by the proceedings Editors. Reviews were conducted by expert referees to the professional and scientific standards expected of a proceedings journal published by IOP Publishing.

IOP Conf. Series: Materials Science and Engineering **378** (2018) 012001 doi:10.1088/1757-899X/378/1/012001

# Temporal double-slit interferences of a quantum fluxon in quantum Josephson nanocircuits

**A Iwamoto and N Hatakenaka***

Graduate School of Integrated Arts and Sciences, Hiroshima University, Higashi-Hiroshima 739-8521, Japan

*E-mail: noriyuki@hiroshima-u.ac.jp

**Abstract.** We investigate temporal Young's double-slit interferences of a quantum fluxon in quantum Josephson nanocircuits, in order to explore the nonlocality of time at macroscopic scales. We employ Gaussian wave-packet approaches to describe the time evolution of the quantum fluxon. We obtained an analytic formula for the probability density of finding the quantum fluxon at a detector's position as a function of time, and found interference fringes in the time domain. This shows nonlocal nature of time at macroscopic scales and also provides potential applications for a new type of building blocks of quantum information in solid state devices.

## 1. Introduction

Quantum mechanics led us to a counterintuitive concept, i.e. wave-particle duality. Young's double slit experiment which revealed the wave nature of the light also greatly contributed to clarify this concept from the viewpoint of wave nature of the matter by running the same setup except a special feature of a source that outputs just one photon at a time. Needless to say, Heisenberg's uncertainty principle lies behind this concept. It has been predominantly studied on quantum nonlocality based on the uncertainty relation between position and momentum. On the other hand, it is possible to study quantum nonlocality based on another type of uncertainty relation, i.e. uncertainty of energy and time. In this case, the nonlocality is related to time. Thus, there has been considerable interest in temporal version of double-slit experiments [1-4] and has been successfully confirmed in experiments at *microscopic* scales.

In contrast, there has been discussed as another fundamental problem in quantum mechanics. It is a question of whether quantum mechanics is applicable to *macroscopic* systems. A concreate example is a superposition of quantum states at macroscopic scales, i.e. the Schrödinger cat as well as double-slit interferences. This issue has been tested in various systems. It is noteworthy that interference fringes were demonstrated in a double slit experiment of a fullerene molecule regarded as a macroscopic object [5]. However, double-slit interference in the time domain has not yet been demonstrated so far.

In this paper, we propose a simple scheme for investigating temporal Young's double-slit interferences to explore quantum nonlocality in *time* at *macroscopic* scales through a quantum fluxon that allows delicate control with state-of-the-art technology instead of fullerene $C_{60}$ used in the previous studies.

IOP Conf. Series: Materials Science and Engineering **378** (2018) 012001 doi:10.1088/1757-899X/378/1/012001

## 2. Temporal Young's double-slit interferences of a quantum fluxon

### *2.1. Quantum fluxons*

A Josephson junction is a superconducting device made up of two superconducting materials separated by thin insulating layer capable of electron pair (Cooper pair) tunneling. The tunnel current across the junction can flow depending on the phase difference of the order parameters between superconductors without an applied voltage. The phase difference is governed by sine-Gordon equation based on the Maxwell equation in a long Josephson junction (see Figure 1).

The sine-Gordon equation is a nonlinear wave equation involving a unique solution called kink (anti-kink). The kink solution expresses a topological nonlinear solitary wave travelling in the nonlinear dispersive system without changing its shape and exhibits particle nature even if it is a basically wave. The physical background of this solution is circulating currents across the junction that produces magnetic field locally. Since it possesses a quantum unit of magnetic flux $\Phi_0 = \hbar/2e$, it was named fluxon. Thus, the fluxon is a nonlinear elementary excitation within a long Josephson junction and behaves like a Newtonian particle moving along the Josephson circuits. Note that it is regarded as a *macroscopic* object since the fluxon is created by a huge number of circulating electrons (supercurrents) across the junction.

The fluxon has been considerably studied in applications to information transfers or controls within the framework of classical physics so far [6]. Later, quantum nature of the fluxon has been demonstrated experimentally in quantum dissociation of a bound fluxon pair in annual Josephson circuits [7]. The quantum-mechanical interference of fluxon has been also considered theoretically in connection of the superposition of spatially distinct positions of a fluxon [8], but it has not yet observed directly due to mainly technical difficulty for making spatial fluxon interferometer in current fabrication technology. In contrast, it seems that a temporal interferometer we discuss below is feasible to achieve at existing technologies.

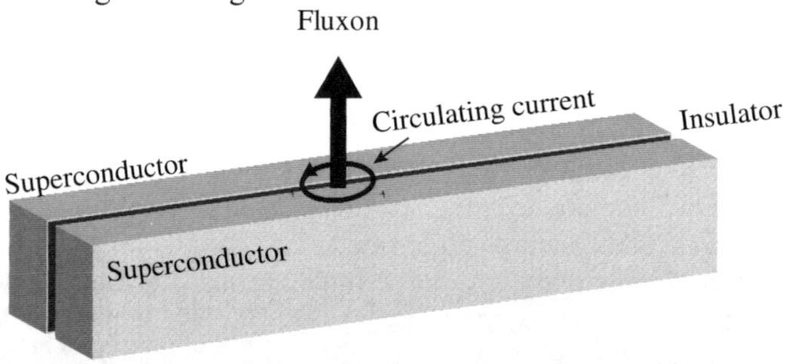


**Figure 1.** A schematic diagram of a fluxon created by a circulating current in a long Josephson junction composed of an insulating thin layer sandwiched by two superconductors.

### *2.2. Temporal Young's double-slit interferometers*

In a temporal double-slit experiment, only one slit with a shutter is placed in front of the screen (or detector) in contrast to a spatial double-slit experiment. The shutter that can be opened and closed forms double slits in different times.

Suppose that a fluxon is initially pinned by a microshort of a long Josephson junction. The fluxon is then irradiated by a coherent superposition of two *temporally* separated superposition states for a single photon, sometimes called a time-bin photon, which is produced by a Mach-Zender interferometer. The pinned fluxon can then move due to collision with the photon. Since the photon is in temporally superposition states, the moving fluxon is also in temporally superposition states. In other words, the superposition of macroscopically distinct states in time is produced by the media conversion from a time-bin photon to a time-bin fluxon. Figure 2 shows the schematic diagram of a setup for temporal Young's double slit experiments in a Josephson system we described above.

IOP Conf. Series: Materials Science and Engineering **378** (2018) 012001 doi:10.1088/1757-899X/378/1/012001

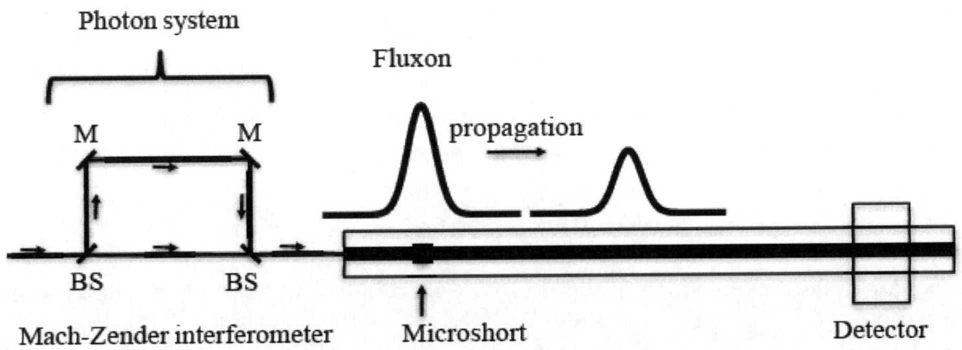


**Figure 2.** A proposed setup of temporal Young's double-slit experiment of a quantum fluxon. BS and M stand for a half beam splitter and a mirror for a photon, respectively.

The trajectories of a fluxon in spacetime for this interferometer are shown in Figure 3. A fluxon is initially prepared at a microshort. At $t = 0$, the fluxon starts to move with a 50% chance due to the collision with an advanced photon of a time-bin photon. The remaining 50% portion of the fluxon remains intact until a retarded photon comes. After the duration time $T_1$, The remaining fluxon will start to move at the same velocity of the advanced portion of the fluxon. We assume that fluxon will eventually move by collision with superposed photons. The interference terms are ignored to avoid complication in Figure 3.

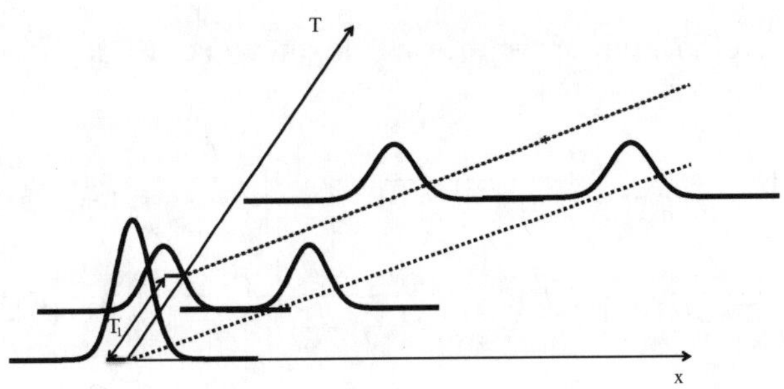


**Figure 3.** Trajectories of a fluxon after collisions with a single photon in a temporally superposition states.

## 3. Time evolution of temporally superposition states of a quantum fluxon

### *3.1. Gaussian wave-packet approach*

Now let us calculate time evolution of the wavefunction in this system. The time evolution of the wavefunction of a free quantum fluxon with mass $m$ is given by

$$\psi(x,t) = \int_{-\infty}^{\infty}\int_{-\infty}^{\infty} \mathrm{e}^{-\frac{\mathrm{i}}{\hbar}\frac{p^2}{2m}t}\mathrm{e}^{-\frac{\mathrm{i}}{\hbar}p(x-x')}\psi(x',0)\mathrm{d}x'\mathrm{d}p \tag{1}$$

where $p$ is a fluxon momentum. We employ Gaussian wave packet as an initial state of a stationary fluxon given as

$$\psi(x,0) = \frac{1}{(2\pi{\sigma_0}^2)^{1/4}} \mathrm{e}^{-\frac{x^2}{4{\sigma_0}^2}} \tag{2}$$

where $\sigma_0$ is a parameter for a Gaussian width. The wave function of the fluxon which start to move by collision with a photon of momentum $p_0$ can be represented as

$$\psi'(x,0) = \psi(x,0)\mathrm{e}^{\frac{\mathrm{i}}{\hbar}p_0 x}. \tag{3}$$

Using this equation, the wavefunction of the moving fluxon in the first part of the superposition state at time $t$ is given by

$$\begin{aligned}\psi_1(x,t) &= \int_{-\infty}^{\infty}\int_{-\infty}^{\infty} \mathrm{e}^{-\frac{\mathrm{i}\,p^2}{\hbar 2m}t}\mathrm{e}^{-\frac{\mathrm{i}}{\hbar}p(x-x')}\psi'(x',0)\mathrm{d}x'\mathrm{d}p \\ &= \left[\frac{1}{\sqrt{2\pi}\sigma_0\left(1+\mathrm{i}\frac{t}{\tau}\right)}\right]^{1/2} \exp\left\{-\frac{(x-v_0 t)^2}{4{\sigma_0}^2\left[1+\left(\frac{t}{\tau}\right)^2\right]}\left(1-\mathrm{i}\frac{t}{\tau}\right)+\frac{\mathrm{i}}{\hbar}\left[p_0 x-\frac{p_0^2}{2m}t\right]\right\}\end{aligned} \tag{4}$$

where $v_0$ is a velocity converted from momentum $p_0$. On the other hand, the rest part of the superposition state of the fluxon remains intact until the subsequent partner photon in the Mach-Zender interferometer comes at time $T_s$, i.e. the time slit interval. This expression is obtained by inserting equation (2) into equation (1) as

$$\psi_2(x,T_s) = \left[\frac{1}{\sqrt{2\pi}\sigma_0\left(1+\mathrm{i}\frac{T_s}{\tau}\right)}\right]^{1/2} \exp\left\{-\frac{x^2}{4{\sigma_0}^2\left[1+\left(\frac{T_s}{\tau}\right)^2\right]}\left(1-\mathrm{i}\frac{T_s}{\tau}\right)\right\}. \tag{5}$$

The expression of the wavefunction for the moving fluxon by collision with the partner photon of the same momentum $p_0$ at time $t$ is given by

$$\begin{aligned}\psi_2(x,t) &= \left[\frac{m}{2\pi\hbar\mathrm{i}(t-T_s)}\right]^{1/2}\int_{-\infty}^{\infty}\mathrm{e}^{-\frac{im}{2\hbar(t-T_s)}(x-x')^2}\left[\psi_2(x',T_s)\mathrm{e}^{\frac{\mathrm{i}}{\hbar}p_0 x}\right]\mathrm{d}x' \\ &= \left[\frac{1}{\sqrt{2\pi}\sigma_0\left(1+\mathrm{i}\frac{t}{\tau}\right)}\right]^{1/2} \exp\left\{-\frac{[x-v_0(t-T_s)]^2}{4{\sigma_0}^2\left[1+\left(\frac{t}{\tau}\right)^2\right]}\left(1-\mathrm{i}\frac{t}{\tau}\right)+\frac{\mathrm{i}}{\hbar}\left[p_0 x-\frac{p_0^2}{2m}(t-T_s)\right]\right\}.\end{aligned} \tag{6}$$

Here we introduce a comoving frame by setting the spatial slit interval as $v_0 T_s$, in order to simplify the situation. Then the wavefunctions can be reexpressed as

$$\psi'_1(x,t) = \left[\frac{1}{\sqrt{2\pi}\sigma_0\left(1+\mathrm{i}\frac{t+T_s}{\tau}\right)}\right]^{1/2} \exp\left\{-\frac{[x-v_0(t+T_s)]^2}{4{\sigma_0}^2\left[1+\left(\frac{t+T_s}{\tau}\right)^2\right]}\left(1-\mathrm{i}\frac{t+T_s}{\tau}\right)+\frac{\mathrm{i}}{\hbar}\left[p_0 x-\frac{p_0^2}{2m}(t+T_s)\right]\right\}, \tag{7}$$

$$\psi'_2(x,t) = \left[\frac{1}{\sqrt{2\pi}\sigma_0\left(1+\mathrm{i}\frac{t+T_s}{\tau}\right)}\right]^{1/2} \exp\left\{-\frac{[x-v_0(t-T_s)]^2}{4{\sigma_0}^2\left[1+\left(\frac{t+T_s}{\tau}\right)^2\right]}\left(1-\mathrm{i}\frac{t+T_s}{\tau}\right)+\frac{\mathrm{i}}{\hbar}\left[p_0 x-\frac{p_0^2}{2m}(t-T_s)\right]\right\}. \tag{8}$$

The whole wavefunction is represented by the superimposition of these,

$$\psi(x,t) = N[\psi'_1(x,t)+\psi'_2(x,t)] \tag{9}$$

where the normalization constant $N$ can be represented as

$$N = \left(2\left\{1 + \exp\left[-\frac{(v_0 T_S)^2}{4{\sigma_0}^2}\right]\cos\left(\frac{p_0^2}{m\hbar}T_s\right)\right\}\right)^{-1/2} . \tag{10}$$

Therefore, the probability density is then given as

$$|\psi(x,t)|^2 = \frac{2N^2}{\sqrt{2\pi}\sigma_0\left[1+\left(\frac{t+T_S}{\tau}\right)^2\right]^{1/2}}\exp\left\{-\frac{2[(x-v_0t)^2+(v_0T_S)^2]}{4{\sigma_0}^2\left[1+\left(\frac{t+T_S}{\tau}\right)^2\right]}\right\}$$

$$\times\left(\cosh\left\{\frac{4(x-v_0t)v_0T_S}{4{\sigma_0}^2\left[1+\left(\frac{t+T_S}{\tau}\right)^2\right]}\right\} + \cos\left\{\frac{4(x-v_0t)v_0T_1}{4{\sigma_0}^2\left[1+\left(\frac{t+T_S}{\tau}\right)^2\right]}\left(\frac{t+T_S}{\tau}\right) + \frac{p_0^2}{m\hbar}T_S\right\}\right) . \tag{11}$$

The hyperbolic cosine term in the parenthesis shows the classical behavior of two Gaussian wave packets, while the cosine term represents interference between them. It is noteworthy that the frequency of interference depends on time. We will discuss this feature concretely in the following.

### 3.2. *Interference fringes*

Now we display the time evolution of the wave packet based on the obtained expression. Here we assume that the mass of fluxon is about 1/100 of the electron [9] and the fluxon speed is 1/100 of the speed of light [10]. The initial spread of the Gaussian function $\sigma_0$ is assumed to be the size of fluxon $\lambda_J = 1 \times 10^{-6}$m. In the following discussion, optimization for the observation is not done. However, experimental observation can be feasible since the time scale is located in picoseconds.

Figure 4 shows the probability density of finding a fluxon as a function of time at various detector positions. Over time, a second wave packet is observed after the first wave packet. In addition, as a feature of the Gaussian wave packet, it can be seen that the amplitude decreases with the wave packet propagation and the wave packet also spreads due to wave number dispersion. The spread of the wave packet leads to the superposition of the two wave packets, and the interference pattern is manifested at the observation point away from the slit where the overlap becomes large.

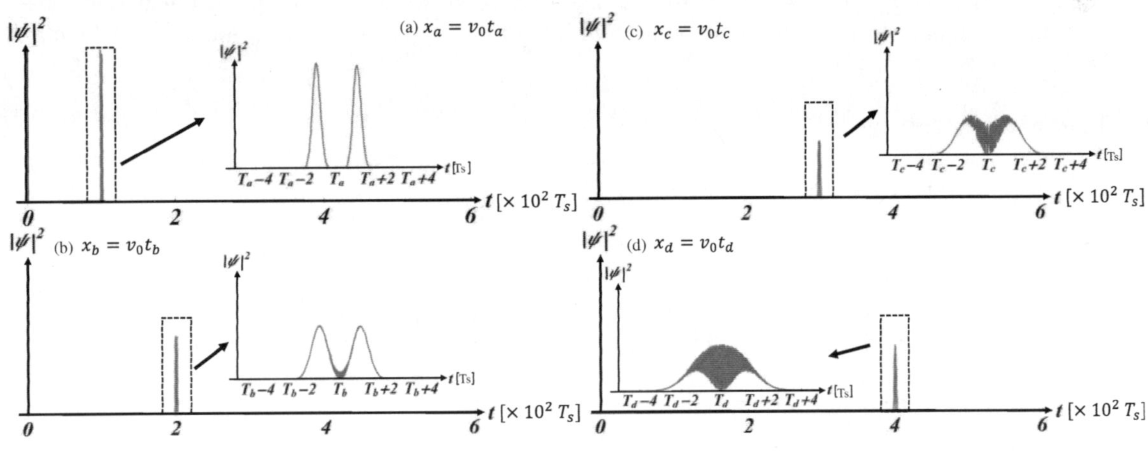


**Figure 4.** Probability density of a fluxon as a function of time at various detector positions, (a) for $10^2$T$_s$ , (b) for $2 \times 10^2$T$_s$, (c) for $3 \times 10^2$T$_s$ and (d) for $4 \times 10^2$T$_s$, respectively.

The characteristics of temporal interference using the Gaussian wave packets appear in the oscillation period. Unlike the oscillation patterns in a space double-slit experiment, the oscillation periods are not same at different detector positions. The oscillation frequencies of interference fringes in $|\psi(x,t)|^2$ is given as

$$\omega(t) = \frac{v_0^2 T_S(1+T_S)}{\sigma_0^2\left\{1+\frac{(t+T_S)^2}{\tau^2}\right\}\tau t} \tag{12}$$

and is shown as a function of time in Figure 5. The frequency monotonically decreases with time. Therefore, the period of the interference pattern increases as the position of the detector moves away from the microshort at which fluxon starts to move. This feature is a strong evidence for temporal interference.

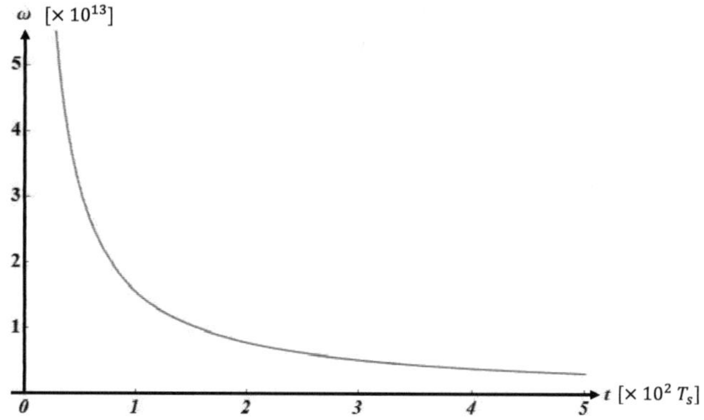


**Figure 5.** Time dependent oscillation frequencies of interference fringes.

## 4. Concluding remarks

The temporal Young's double-slit interferences of a quantum fluxon have been investigated for exploring quantum nonlocality of time at *macroscopic* scales. We employ Gaussian wave-packet approaches to describe the time evolution of the quantum fluxon. We have obtained an analytic formula for the probability density of finding the quantum fluxon at a detector's position as a function of time and found interference fringes in the time domain with a frequency dependent on the detector position. This showed nonlocal nature of time at a macroscopic scale. This temporal interference also provides potential applications for a new type of building blocks of quantum information in solid state devices.

## 5. Acknowledgements

We would like to thank S Ishizaka for helpful discussions and suggestions. This work was supported in part by JSPS KAKENHI Grant Number JP17K05579.

## 6. References

[1] Szriftgiser P, et al. 1996 *Phys. Rev. Lett.* **77** 4
[2] Lindner F 2005 *Phys. Rev. Lett.* **95** 040401
[3] Ishikawa K L 2006 *Phys. Rev. A* **74** 023806
[4] Bauer M 2014 *Am. J. Phys.* **82** 1087
[5] Arndt M, et al. 1999 *Nature* **401** 680
[6] Costabile G, et al. 1978 *Appl. Phys. Lett.* **32** 5877
[7] Fistul M V 2003 *Phys. Rev. Lett.* **91** 257004
[8] Shnirman A, et al. 1995 *Phys. Rev. B* **52** 3541
[9] Kato T and Imada M 1996 *J. Phys. Soc. Jpn.* **65** 2963
[10] Swihart J C 1961 *J. Appl. Phys.* **32** 461

IOP Conf. Series: Materials Science and Engineering **378** (2018) 012002 doi:10.1088/1757-899X/378/1/012002

# The electrochromic property of $MoO_3/WO_3$ nanocomposite films

**Surachai Chaichana[1,a], Lek Sikong[1,2,b], Kalayanee Kooptarnond[1,c], and Kanadit Chetpattananondh[3,d]**

[1]Department of Mining and Materials Engineering, Faculty of Engineering, Prince of Songkla University, Hat Yai, Songkhla, 90112, Thailand

[2]Center of Excellence in Materials Engineering (CEME), Prince of Songkla University, Hat Yai, Songkhla, 90112, Thailand

[3]Department of Electrical Engineering, Faculty of Engineering, Prince of Songkla University, Hat Yai, Songkhla, 90112, Thailand

E-mail: [a]asuraguy_aunt@hotmail.com, [b]lek.s@psu.ac.th, [c]kalayanee.k@psu.ac.th, [d]kanadit.c@psu.ac.th

**Abstract.** Electrochromic films of molybdenum / tungsten trioxide ($MoO_3/WO_3$) prepared from a solution of peroxotungstic acid (PTA) were formed on indium doped tin oxide (ITO) glass substrate using the sol-gel and dip coating methods at difference ratios of $MoO_3$ of 15, 30 and 50 %mol. The effect of $MoO_3$ on the structure, morphology, optical and electrochromic properties of films were investigated by using X-ray diffraction (XRD), X-ray photoelectron spectroscopy (XPS), Atomic force microscope (AFM), Scanning electron microscope (SEM), UV-vis spectrophotometer and Cyclic voltammogram (CV), respectively. The 30 %mol $MoO_3$ doped $WO_3$ film demonstrated a good electrochemical properties in terms of diffusion coefficient better than other ratios and pure $WO_3$ due to its large surface area of the high roughness composite film.

## 1. Introduction

Currently, the protection of heat waves through the glass windows, the electrochromic film as a smart window [1] is mostly useful to reduce heat in buildings. This type of glass consists of a layer of film bounded by metal oxide layers. The more special feature is the ability to modify the optical properties

IOP Conf. Series: Materials Science and Engineering **378** (2018) 012002 doi:10.1088/1757-899X/378/1/012002

by supplying electricity to the film system, which can be transformed from translucent glass into darker or more opaque glass and can be returned to the translucent one with low electric current. It also controls the amount of light transmittance. Electrochromic phenomenon materials containing properties of heat protection shine through the glass, using semiconductor metal oxide film coat on glass are many types of such as $TiO_2$, CrO, $Nb_2O_5$, $SnO_2$, NiO, $IrO_2$ [2], $WO_3$, and $MoO_3$ [3,4]. Researchers have different methods of forming as sputtering [5], chemical vapor deposition (APCVD) [6], dipping [7] and sol-gel method [1,4,8]. Among these techniques, sol-gel method is a simple, low cost method that does not require electrical appliances and it can control film quality.

In this research, the effect of $MoO_3$ in electrochromic layer $WO_3$ films with high concentration 15-50 % mol on optical and electrochromic properties of the films were investigated. The films coated on ITO glass substrates were prepared by sol-gel and dip coating methods and they were characterized and determined electrochromic and optical properties using the sophisticated techniques.

## 2. Experimental

### *2.1. Preparation sol-gel*

#### *2.1.1. Synthesis of $WO_3$ sol*

The $WO_3$ sol prepared from tungsten powder (6.5 g ; 99.9%, Merck) dissolved into 40 ml of 35% hydrogen peroxide and 4 ml of distilled water that will be got PTA precursor solution [9] and sol dip coating was prepared by mixing 10 ml of PTA solution and 10 ml of ethanol for 30 min stirring. Finally, the $WO_3$ sol was kept at a constant temperature before coating on ITO glass substrate.

#### *2.1.2. Synthesis of $MoO_3/WO_3$ composite sol*

Firstly, Ammonium heptamolybdate tetrahydrate or AHT ($(NH_4)_6Mo_7O_{24}\cdot 4H_2O$, Ajax Finechem) powder was dissolved in distilled water at a 60 °C under magnetic stirring [10]. Then adjusted the solution pH~4 by adding nitric acid ($HNO_3$). The colorless solution was obtained and sol dip coating was prepared by mixing 10 ml of the AHT solution and 10 ml of ethanol with stirring at room temperature. After that the mixed solution was added into $MoO_3$ sol with concentration of 15, 30 and 50 %mol at room temperature under magnetic stirring then the pale-yellow solution was formed. Finally, it was kept at a constant temperature for 24 h before coating on ITO glass substrate.

### *2.2. Preparation electrochromic films of $MoO_3/WO_3$*

Dip coating was applied for preparation electrochromic films of $MoO_3/WO_3$ on ITO glass substrate (2.5 x 2.5 cm). Firstly, ITO glass was cleaned with distilled water, acetone ($CH_3COOH$) and ethanol ($C_2H_5OH$), respectively. Thin film dip coatings were performed using pure $WO_3$ sol, 15, 30, 50 %mol $MoO_3/WO_3$ composite sol and pure $MoO_3$ sol and the prepared films were dried at room temperature. All films were annealed at 300 °C for 2 h at heating rate of 5 °C/min.

### *2.3. Analysis techniques*

X-ray diffraction (XRD, X'Pert MPD, Philips) was performed by a diffractometer with CuKα radiation to characterize the film microstructures after calcined at a temperature 300 °C. The surface morphology and cross section image of the film was analyzed using (SEM, Quata400). X-ray photo electron spectroscopy (XPS; AXIS ULTRA$^{DLD}$, Kratos analytical, Manchester UK.) was used to confirm valence states of Mo and W in film while surface roughness was determined using an atomic force microscopy (AFM). The electrochromic property of the film was done in terms of electrochemical cyclic voltammetry (CV) using Pt and Ag/AgCl as the working, counter and reference electrodes in 0.5 M sulfuric acid ($H_2SO_4$) electrolyte. The optical absorption and transmittance spectra were measured by a UV–vis spectrophotometer (GENESYS 10S, Thermo fisher scientific) with wavelength range of 350-750 nm.

## 3. Results and discussion

### *3.1. Structural properties*

The film compositions of $WO_3$, 15, 30 and 50 %mol $MoO_3/WO_3$ were analyzed by the XRD diffraction shown in figure 1-a. It was found that the structure of sample is monoclinic $WO_3$ crystal phase at 23.2°, 24.3° and 28.3° [11] while hexagonal $MoO_3$ (h-$MoO_3$) was formed at 16.9°and 25.8° for 15 and 30% mol of $MoO_3/WO_3$ films [12]. However, a small peak of $MoO_3$ at 19.4° [12] and amorphous phases of $MoO_3$ and $WO_3$ were formed for 50 %mol $MoO_3/WO_3$. Also, at the 2Θ (degree) crystalline structures of ITO glass presented peaks identified is aspect analogous [9]. Besides, 30 %mol $MoO_3/WO_3$ films were confirmed oxidation states of molybdenum and tungsten by using XPS technique (figure 1-b) that showed spectra of tungsten (W) at 36.19 eV (W $4f_{7/2}$) and 38.34 eV (W $4f_{5/2}$) of $WO_3$. Also, Figure 1-c demonstrated spectra of molybdenum (Mo) at 233.22 eV (Mo $3d_{5/2}$) and 336.39 eV (Mo $3d_{3/2}$) of $MoO_3$ on the ITO glass substrate.

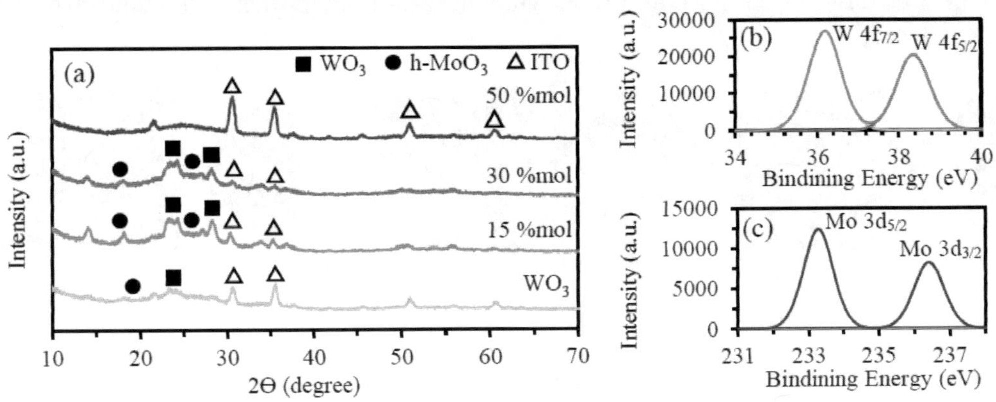


**Figure 1.** The XRD analysis of $WO_3$, 15, 30 and 50 %mol $MoO_3/WO_3$ films (a) and XPS analysis of 30 %mol $MoO_3/WO_3$ film showing spectra of W and Mo, respectively (b,c).

*3.2. Morphology study*

SEM image of $WO_3$, 30 %mol $MoO_3/WO_3$ and $MoO_3$ films coated on ITO glass substrate are shown in figure 2. All films are uniformly distributed but agglomeration of $MoO_3$ particles (figure 2-c) had larger than the $WO_3$ and 30 %mol $MoO_3/WO_3$ films (figure 2-a,e) while fig. 2-b,d,f showed X-ray mapping of W and Mo of film layer coated on ITO. The EDX mapping confirmed that $MoO_3$ and $WO_3$ of the 30 %mol $MoO_3/WO_3$ film were well dispersion and contained approximately 29.4% W and 5.3% Mo. (figure 2-g).

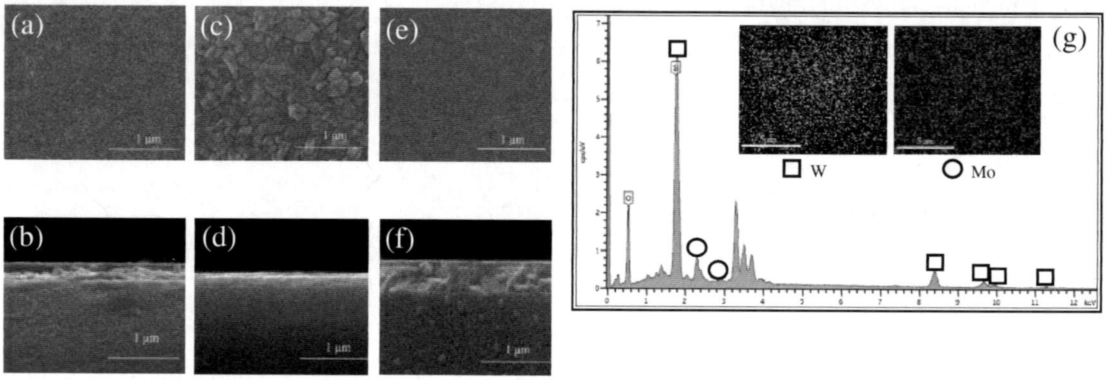


**Figure 2.** The SEM at high magnification 50,000x the annealed films of $WO_3$ (a,b), $MoO_3$ (c,d), 30 %mol $MoO_3/WO_3$ (e,f) and EDX mapping of 30 %mol $MoO_3/WO_3$ film (g).

As $MoO_3$ was added to in $WO_3$, the surface roughness of the composite films increased as compared to pure $WO_3$ for instance, 30 %mol $MoO_3/WO_3$ film increased to 3.2 nm from 1.6 nm for pure $WO_3$ while it was 3.6 nm for pure $MoO_3$. That resulted in an increase in surface area of the film which led to high electrochromic performance. The thickness of the films were approximately 90-170 nm from AFM (figure 3) and films were smooth surfaces with thickness from SEM cross-section analysis about 100-200 nm (Figure 2-f).

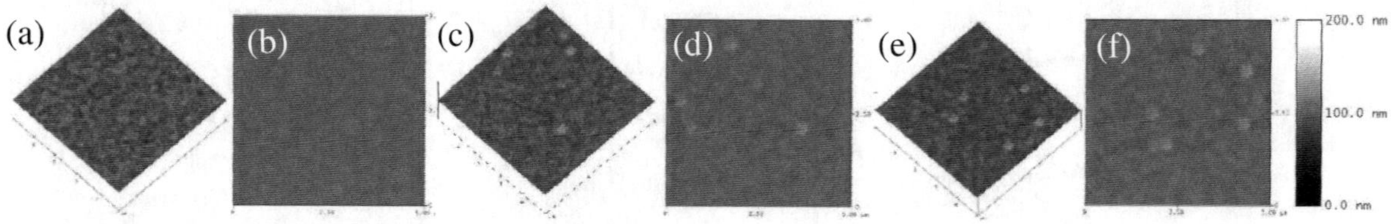


**Figure 3.** The 3D and 2D AFM images of $WO_3$ (a,b), 30 %mol $MoO_3/WO_3$ (c,d) and $MoO_3$ (e,f) films coated on ITO substrates.

*3.3. Electrochromic properties*

The transmittance of for $WO_3$, 15, 30 and 50 %mol $MoO_3/WO_3$ films were analyzed using the UV-vis analysis with spectra wavelength in visible light range (350-750 nm) exhibited in the table 1 and figure 4-a. The color modulation in terms of difference of transmittance between bleached and colored states and charges diffusion coefficients of the $WO_3$ and $MoO_3/WO_3$ nanocomposite films were determined by cyclic voltammeter under applied potential of at 1.0 V and -1.0 V in $H_2SO_4$ 0.5 M solution also shown in table 1. The 30 %mol $MoO_3/WO_3$ exhibited the highest color difference of bleached and colored states (44.04%) more than those of $WO_3$, 15 and 50 %mol $MoO_3/WO_3$ films (31-32%) and the same with the diffusion coefficient was highest since it had most surface roughness and surface area.

**Table 1.** Optical and electrochromic properties of the synthesized films

| Samples | Transmittance (%) | | | Diffusion coefficient ($cm^2/s$) |
|---|---|---|---|---|
| | bleached, $T_b$ | colored, $T_c$ | $\Delta T$ | |
| Pure $WO_3$ | 85.83 | 54.26 | 31.57 | $2.71 \times 10^{-11}$ |
| 15 %mol $MoO_3/WO_3$ | 74.79 | 41.91 | 32.88 | $2.91 \times 10^{-11}$ |
| 30 %mol $MoO_3/WO_3$ | 90.25 | 46.21 | 44.04 | $3.35 \times 10^{-11}$ |
| 50 %mol $MoO_3/WO_3$ | 86.94 | 54.31 | 32.63 | $2.42 \times 10^{-11}$ |

Moreover, The electrochromic film with electrical properties of $WO_3$, 15, 30 and 50 %mol $MoO_3/WO_3$ films were measured by using CV in 0.5 M $H_2SO_4$ solution at a potential scan rate of 100 mV/s from 1.0 V to -1.0 V shown in figure 4-b. Their diffusion coefficients were calculated from CV analysis to indicate the electrochromic performance of the films summarized in table 1. It was apparent that 30 %mol $MoO_3/WO_3$ film exhibited the maximum diffusion coefficient about $3.35 \times 10^{-11}$ $cm^2/s$ while those of pure $WO_3$, 15 and 50 %mol $MoO_3/WO_3$ were $2.71 \times 10^{-11}$, $2.91 \times 10^{-11}$ and $2.42 \times 10^{-11}$ $cm^2/s$, respectively. This value of 50 %mol $MoO_3/WO_3$ was less than the other films because it is probably its amorphous phases of $MoO_3$ and $WO_3$ as mentioned by XRD analysis in figure 1-a. Particle dispersion consistently, smooth surface (or less surface area) perhaps to affect the electrochromic behavior.

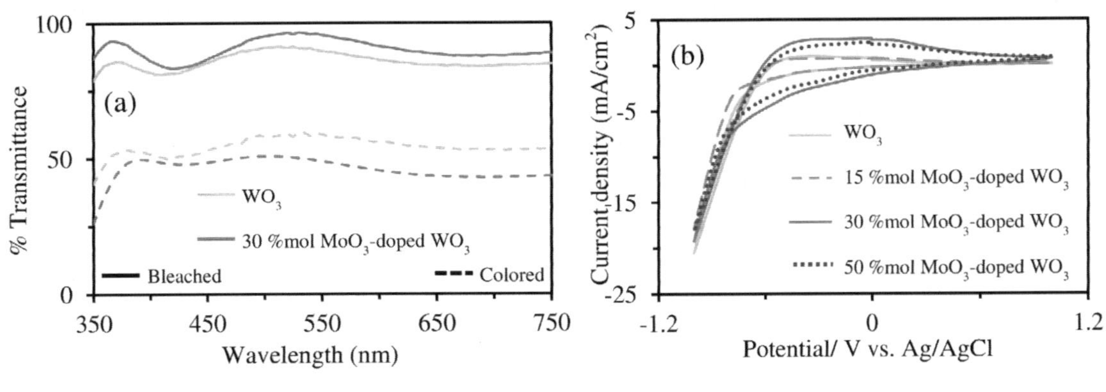

**Figure 4.** Transmittance (%) of the pure $WO_3$ and $MoO_3/WO_3$ (a) and the CV analysis of $WO_3$, 15, 30 and 50 %mol $MoO_3/WO_3$ nanocomposite films measured at 1.0 V to -1.0 V in 0.5M $H_2SO_4$ solution (b).

## 4. Conclusions

The electrochromic of $MoO_3/WO_3$ nanocomposite films were successfully prepared by sol-gel method and formed by dip coating on ITO glass substrates. All films were dispersion consistently, structural analysis of $WO_3$, 15, 30 and 50 %mol $MoO_3$ when $MoO_3$ was added in to $WO_3$ films revealed monoclinic $WO_3$ and h-$MoO_3$ crystalline structures. Higher concentration of $MoO_3$ addition in $WO_3$ such as 50 %mol inhibited growth of $WO_3$ and h-$MoO_3$ of the film and amorphous phase was formed leading to less electrochromic performance while 30 %mol $MoO_3/WO_3$ exhibited the highest diffusion coefficient about 3.35 x $10^{-11}$ $cm^2/s$ and the highest color modulation of 44.04% due to its high surface roughness or surface area. Therefore, this film shows high electrochromic property.

## 5. Acknowledgements

The author would like to thank the Center of Excellence in Materials Engineering (CEME) and Faculty of Engineering, Prince of Songkla University for expenses support. Moreover the authors would like to express my sincere thanks to Asst. Prof. Dr. Pipat Chooto for cyclic voltammetry analyses.

## 6. References

[1] P. P. Gonzalez-Borrero, F. Sato, A.N. Medina, M.L. Baesso and A.C. Bento: Applied Physics Letters. Vol. 96, 061909 (2010); doi: http://dx.doi.org/10.1063/1.3313945

[2] J. Livage and D. Ganguli: Solar Energy Materials & Solar Cells. Vol. 68 (2001), p. 365-381

[3] C.S. Hsu, C.C. Chan, H.T. Huang, C.H Peng and W.C Hsu: Thin Solid Films. Vol. 516 (2008), p. 4839-4844

[4] S.Y. Lin, C.M. Wang, K.S. Kao, Y.C. Chen and C.C. Liu: Journal of Sol Gel Science and Technology. Vol. 53 (2010), p. 51

[5] V. Madhavi, P. Kondaiah, O.M. Hussain and S. Uthanna: Conference paper in energy. Vol. 1 (2013), p. 1-5

[6] T. Ivanova, K.A. Gesheva, M. Kalitzova, F. Hamelmann, F. Luekermann and U. Heinzmann: Journal of Optoelectronics and Advanced Materials. Vol. 11 (2009), p. 1513-1516

[7] A. Novinrooz, M. Sharbatdaran and H. Noorkojouri: Central European Science Journals (2005), p. 456-466

[8] M. Sharbatdaran, A. Novinrooz and H. Noorkojouri: Iranian Journal of Chemistry and Chemical Engineering. Vol. 25 (2006), p. 25-29

[9] E. Bica, M. Stefan, L.E. Muresan, E. Indrea, I.C. Popescu and E.J. Popovici: Journal of Optoelectronics and Advanced Materials Symposia. Vol. 1 (2009), p. 1011-1014

[10] P. Jittiarporn, L. Sikong, K. Kooptarnond and W. Taweepreda: Ceramics International. Vol. 40 (2014), p. 13487-13495

[11] K. Paipitak, W. Techitdheera, S. Porntheeraphat and W. Pecharapa: Energy Procedia. Vol. 34 (2013), p. 689-696

[12] P. Jittiarporn, L. Sikong, K. Kooptarnond, W. Taweepreda, P. Chooto, M. Khangkhamano: Applied Mechanics and Materials.Vol.835 (2016), p. 34-41

IOP Conf. Series: Materials Science and Engineering **378** (2018) 012003 doi:10.1088/1757-899X/378/1/012003

# Black sesame-derived carbon dots for metal ion and amine vapour sensing

**Peerasak Paoprasert[1*], Hataipat Moonmuang[1], Preeyanuch Supchocksoonthorn,[1] Nichaphat Thongsai,[1] and Sumana Kladsomboon[2]**

[1]Department of Chemistry, Faculty of Science and Technology, Thammasat University, Pathumthani, 12120 Thailand

[2]Center for Research and Innovation, Faculty of Medical Technology, Mahidol University, Nakhon Pathom, 73170 Thailand

*Corresponding author, email: peerasak@tu.ac.th

**Abstract.** Carbon dots (CDs) are carbon nanoparticles with unique properties, including high photo stability and low toxicity. In this work, CDs were synthesized from black sesame via one-step hydrothermal method. CDs were characterized using UV-visible spectroscopy, fluorescence spectroscopy, Fourier-transform infrared spectroscopy, transmission electron spectroscopy, and x-ray photoelectron spectroscopy. The results showed that the size of CDs was around 7 nm, and they exhibited blue fluorescent emission with a quantum yield of 1%. The obtained CDs were applied in sensing applications for metal ion detection and ethylenediamine vapour using an electronic nose. This work thus demonstrated that carbon dots prepared from black sesame are intriguing sensing materials for various chemicals.

## 1. Introduction

Xu and coworkers discovered a novel spherical carbon-based nanomaterial In 2004,, which was isolated from an eletrophoretic method for the purification of single-walled carbon nanotubes derived from arc-discharge soot [1]. Carbon dots are considered as emerging luminescent nanoparticles for various applications due to their low toxicity [2], excellent photostability [3], high water solubility [4] and can be prepared by several methods such as, laser ablation [5], pyrolysis [6], hydrothermal[7], and microwave irradiation [8]. Because of these unique and attractive properties, carbon dots have been used in various applications such as sensing [9, 10], electronics [11], and biomedical applications [12, 13].

In this work, a method for synthesizing carbon-dots form black sesame was developed using a one-step hydrothermal method. They were then characterized using UV-visible spectroscopy, fluorescence spectroscopy, Fourier transform infrared (FT-IR) spectroscopy, x-ray photoelectron spectroscopy (XPS), and transmission electron microscopy (TEM). They were used to detect metal ions in aqueous

IOP Conf. Series: Materials Science and Engineering **378** (2018) 012003 doi:10.1088/1757-899X/378/1/012003

solutions and amine vapour, including ammonia, ethylenediamine, and triethyleamine using electronic nose system.

## 2. Experimental section

### *2.1. Materials*

Black sesame was purchased from a local supermarket. All chemicals were purchased from Sigma-Aldrich. Cellulose dialysis membrane (1000 Da MWCO) was bought from Spectrum Labs. Deionized (DI) water was used throughout the experiment.

### *2.2. Methods*

#### *2.2.1. Synthesis of carbon dots*

Black sesame (4 g) and 0.5 M sulfuric acid solution (50 mL) were mixed and reacted at 200 °C for 4 h in a Teflon-lined autoclave. After the reaction completed, the autoclave was cooled down to room temperature. The solution was filtered, neutralized with 10 M NaOH, and centrifuged at 10,000 rpm for 15 min. The CD solution were further purified using dialysis membrane (1,000 Da MWCO) for 72 h, the CD solution was freeze dried to yield brown solid product (0.15 g, 4%)

#### *2.2.2. Quantum yield*

Quantum yield (Φ) of carbon dots was measured using fluorescence spectrometry. Quinine sulfate was used as a reference. The carbon dots were dissolved in DI water (n = 1.33). Quinine sulfate was dissolved in 0.1 M $H_2SO_4$ solution (quantum yield of 54%, n = 1.33). Quantum yield was calculated using equation (1):

$$\Phi = \Phi_R \times \frac{I}{I_R} \times \frac{A_R}{A} \times \frac{n^2}{n_R^2} \quad (1)$$

where I is the integrated emission intensity, A refers to the absorbance, and n is refractive index of the solvent. The subscript R denotes the reference sample.

#### *2.2.3. Electronic nose measurement*

The optical electronic nose system consists of light sources, sensing materials, and a photo detector. Light-emitting diodes (LEDs) and a commercial photo detector (ET-TCS230) were used to detect light intensity. The eight gas sensor arrays were generated based on eight colors of LEDs. The transmitting light intensity through sensing thin film was observed in the form of photon frequency (Hz), captured by a photo-detector.

## 3. Result and discussion

### *3.1. Synthesis and characterization of the carbon dots*

In this work, the carbon dots was prepared from black sesame from a one-step hydrothermal reaction and used for metal ion and amine vapour sensing applications (Figure 1a). Black sesame was chosen

IOP Conf. Series: Materials Science and Engineering **378** (2018) 012003 doi:10.1088/1757-899X/378/1/012003

as it is cheap and readily available raw material. After hydrothermal treatment and purification, the black sesame-derived carbon dots were characterized. The functional groups of carbon dots were identified using FT-IR. The FT-IR spectra indicated the characteristic absorption bands of O–H at 3372 $cm^{-1}$ and C–H stretching vibrations of amine groups at 2963 $cm^{-1}$, C=O stretching vibrations located at 1648 $cm^{-1}$, and C-O stretching vibration at 1137 $cm^{-1}$ (Figure 1b). This supports that carbon dots are highly hydrophilic and water soluble. XPS spectrum showed signature signals of carbon and nitrogen atoms (Figure 2a). Size and morphology of carbon dots were determined by TEM. The TEM image showed spherical carbon dots with a diameter of about 7 nm (Figure 2b). The optical properties of carbon dots were characterized using UV-visible and fluorescence spectroscopy. The UV–visible spectrum exhibited a broad absorption band from 250 to 500 nm (Figure 2c). The absorption centered at 270 nm, corresponding to π-π* transition of the aromatic $sp^2$ domains [14]. The emission was found to be excitation wavelength dependence, showing the highest emission intensity at 380 nm excitation wavelength (Figure 2d). The carbon dots showed blue emission under UV illumination with a quantum yield of 1%.

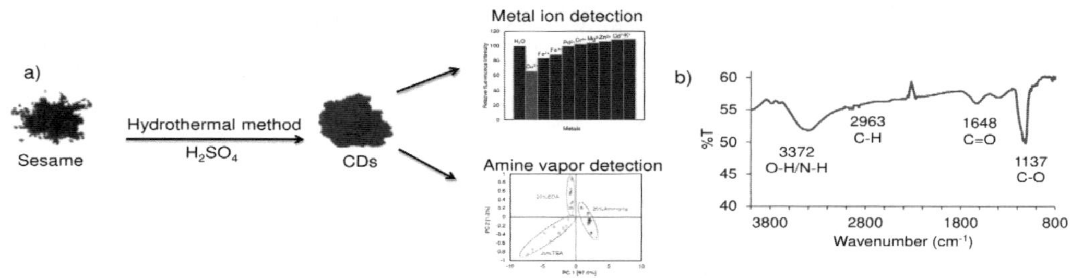


**Figure 1.** (a) Schematic representation of this work and (b) FTIR spectrum of carbon dots

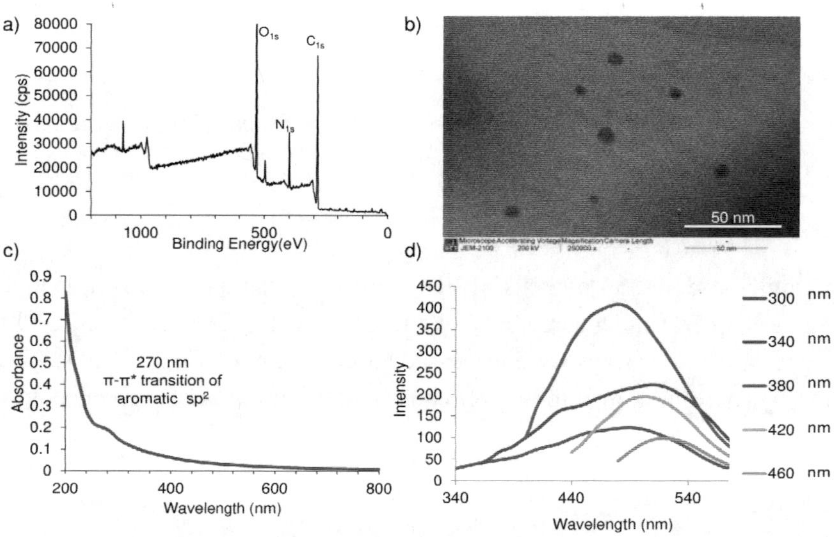


**Figure 2.** (a) XPS survey spectrum of carbon dots, (b) TEM image, (c) UV-vis spectrum, and (d) fluorescent emission spectra at different excitation wavelengths

### *3.2. Sensing applications of the carbon dots*

#### *3.2.1. Metal ion detection*

Carbon dots are known as excellent fluorophores for metal and chemical detection. Carbon dots were applied for the detection of metal ions in solutions, including $Cu^{2+}$, $Fe^{2+}$, $Fe^{3+}$, $Pd^{2+}$, $Cr^{3+}$, $Mg^{2+}$, $Zn^{2+}$, $Cd^{2+}$, and $K^{+}$ (Figure 3a and b). It was found that the carbon dots were the most sensitive to $Cu^{2+}$, possible because the presence of amino groups in carbon dots are known as good ligands for copper ions. This result suggests that our carbon dots are potential sensors for metal ion detection.

#### *3.2.2. Amine vapour sensing*

Carbon dots were used to detect amine vapours, including ammonia, triethyleamine (TEA), and ethylenediamine (EDA) using an electronic nose (Figure 3c). The dynamic optical responses of the film (Figure 3d) corresponded to a change in the light transmission intensity. We found that light transmission of carbon dots in the presence of three different amines were different. This suggested that the carbon dots were successfully tested as an amine vapour sensor. The sensitivity measurement will be done as future work.

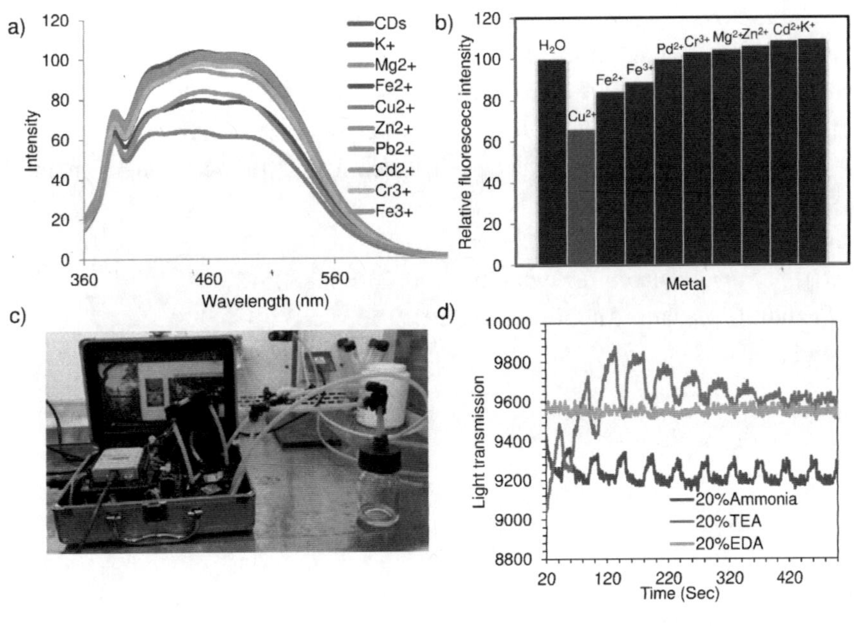


**Figure 3.** (a) Fluorescent spectra of carbon dots in the presence of metal ions, (b) relative fluorescent intensity, (c) electronic nose setup, and (d) gas sensing responses of carbon dots in the presence of amine vapours.

## 4. Conclusion

From this study, carbon dots were prepared from black using a simple hydrothermal method. They are hydrophilic with good solubility in water and showed excitation-dependence fluorescence with a quantum yield of 1%. The carbon dots have been successfully applied for metal ion and amine vapour sensing. They were sensitive to $Cu^{2+}$, showing the highest fluorescent quenching among the metal ions tested in this study. Furthermore, they were able to differentiate three volatile amine compounds, including ammonia, ethylenediamine, and trimethylamine. Future work includes limit of detection measurement and applications in biomedical applications. This work thus demonstrated that carbon dots can be prepared from readily available national precursors and used for various applications.

## 5. Acknowledgement

This research is supported by the governmental budget through Thammasat University in the fiscal year 2017 (Contract No. 6/2560).

## 6. References

[1] Xu, X., et al., *J Am Chem Soc*, 2004. **126**(40): p. 12736-7.
[2] Yang, S.T., et al., *J Phys Chem C Nanomater Interfaces*, 2009. **113**(42): p. 18110-4.
[3] Baker, S.N. and G.A. Baker, *Angew Chem Int Ed Engl*, 2010. **49**(38): p. 6726-44.
[4] Qu, S., et al., *Angew Chem Int Ed Engl*, 2012. **51**(49): p. 12215-8.
[5] Goncalves, H., et al., *Sensors and Actuators B-Chemical*, 2010. **145**(2): p. 702-7.
[6] Zhu, L.L., et al., *Journal of Materials Chemistry C*, 2013. **1**(32): p. 4925-32.
[7] Choi, Y., et al., J. Ind. Eng. Chem., 2017. **47**: p. 329-335.
[8] Zhai, X., et al., *Chem Commun (Camb)*, 2012. **48**(64): p. 7955-7.
[9] Yan, F., et al., *Sensors and Actuators B: Chemical*, 2014. **192**: p. 488-95.
[10] Thongsai, N., et al. Sensors and Actuators B: Chemical, 2017. **253**(Supplement C): p. 1026-1033.
[11] Han, X., et al., *Dalton Trans*, 2013. **42**(29): p. 10380-3.
[12] Sahu, S., et al., *Chem Commun (Camb)*, 2012. **48**(70): p. 8835-7.
[13] Khakbaz, F. and M. Mahani. Anal Biochem, 2017. **523**: p. 32-38.
[14] Wang, Y. and A. Hu, *Journal of Materials Chemistry C*, 2014. **2**(34): p. 6921-39.

IOP Conf. Series: Materials Science and Engineering **378** (2018) 012004 doi:10.1088/1757-899X/378/1/012004

# Ising model with four-spin interaction on honeycomb lattice

**Yuming Yang**[1,2,a] **and Baohua Teng**[1,b]
[1]Department of Applied Physics, University of Electronic Science and Technology of China, Chengdu, 610054, China
[2]School of Mathematical Sciences, University of Electronic Science and Technology of China, Chengdu, 610054, China

[a]yangym@uestc.edu.cn, [b]phytbh@uestc.edu.cn

**Abstract.** The Ising model with the nearest-neighbor interaction $J_2$ and four-spin interaction $J_4$ on the honeycomb-lattice is investigated. The correlation relationships among two-spins and four-spins are calculated. The effects of spin-spin interactions on phase diagrams are investigated analytically and numerically. Evidence from our derivation indicates that the classification of the phase-transition depends on the ratio of $J_4$ to $J_2$.

## 1. Introduction

One of the most interesting phenomena in magnetic materials is ferromagnetism [1-4]. Ising model is a simplest and most popular model of a lattice system, and it can be used to simulate the structure of a ferromagnetic materials. The model and its various variants are extensively studied in statistical physics. Ising model on honeycomb lattice is such an extended application [2],[5-7], which can be used to mimic the material that has a geometry structure of honeycomb type, e.g., graphene[8], honeycomb type ZnO nanostructures [6], etc.

The influence of four-spin interactions on critical properties of Ising model has been studied [1-2],[5],[9-14], e.g. the results of effective-field theory, Monte Carlo simulations, Green's function, and so on. The models with four-spin interactions have been used to describe the material properties of $PbHPO_4$, $H_2C_2O_4$ and $La_2CuO_4$, etc. The phase diagram of ferroelectric thin films based on transverse Ising model with a four-spin interaction has been studied and the ferroelectric feature of the phase diagrams been discussed in Ref. [14].

For the honeycomb lattice Ising model, the thermal behaviours and susceptibility and other properties of the lattice system have been discussed in recent research papers [5],[15-17]. Some rigorous solutions have been obtained in the case of the special situation, e.g. the mixed spin transverse Ising model, a honeycomb lattice with interlayer coupling. However, there are few discussions of four-spin interactions on honeycomb lattice Ising model, though it is an important lattice structure.

The purpose of this work is to investigate the properties of the honeycomb lattice Ising model with the nearest-neighbour interaction $J_2$ and four-spin interaction $J_4$. It is to be hoped that one can obtain an

IOP Conf. Series: Materials Science and Engineering **378** (2018) 012004 doi:10.1088/1757-899X/378/1/012004

understanding of the basic qualitative and quantitative features for the effects of four-spin interactions on this lattice system.

## 2. The model and derivation

The system is an array of $N$ fixed lattice sites that forms a two-dimensional periodic lattice. The geometrical structure of the lattice is honeycomb lattice. $s_i$ ($i$=1,⋯,$N$) associated with lattice site $i$ is a spin variable and its value is a number that is either +1 or -1. The Hamiltonian of the model is given by

$$H = -J_2 \sum_{<ij>} s_i s_j - J_4 \sum_{<ijkl>} s_i s_j s_k s_l \tag{1}$$

where <$ij$> denotes a nearest neighbor ($NN$) pair of spins, <$ijkl$> a quartet of spins (four-spin), i.e., a spin and its three nearest neighbors (see figure 1). The strengths of interaction $J_2, J_4$ are all positive.

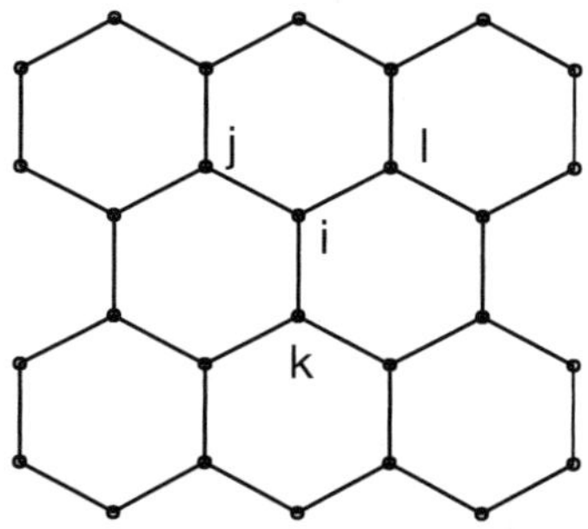


**Figure 1.** Site $i$ and its neighbors. <$ij$>,<$il$> and <$ik$> are nearest neighbors. <$ijkl$> is a quartet of spins and its central spin is site $i$.

To solve this model, let $N_u$ be total numbers of up spins and $N_d$ for total number of down spins. The subscripts of a variable stands for the state of the spin, $u$ for spin-up state and $d$ is just the opposite. $N_{uu}$, $N_{ud}$, $N_{dd}$ are introduced for numbers of $NN$ pair $(1,1)$, $(1,-1)$, and $(-1,-1)$ respectively, where $(1,-1)$ is not distinguished from $(-1,1)$. The variables $N_{uuuu}$, $N_{uuud}$, $N_{uudd}$, $N_{uddd}$, $N_{duuu}$, $N_{duud}$, $N_{dudd}$ and $N_{dddd}$ are introduced for numbers of quartet of spins $(1,1,1,1)$, $(1,1,1,-1)$, $(1,1,-1,-1)$, $(1,-1,-1,-1)$, $(-1,1,1,1)$, $(-1,1,1,-1)$, $(-1,1,-1,-1)$ and $(-1,-1,-1,-1)$ respectively. There are eight types of quartet of spins $(1,1,1,1)$, $(1,1,1,-1)$, $(1,1,-1,-1)$, $(1,-1,-1,-1)$, $(-1,1,1,1)$, $(-1,1,1,-1)$, $(-1,1,-1,-1)$ and $(-1,-1,-1,-1)$, The first item of a four-spin stands for the state of the central spin. Therefore, a state of a four-spin depends on the state of its central spin together with the up-spin or down-spin number of the states of the rest three spins, e.g. $(1,1,1,-1)$ is not distinguished from $(1,1,-1,1)$ or $(1,-1,1,1)$. However, $(1,1,1,-1)$ is distinguished from $(-1,1,1,1)$.

Suppose that the spatial distribution of spin is uniformly at random in a honeycomb lattice system, but the values of $N_u$, $N_d$, $N_{ud}$, $N_{dd}$, $N_{uuuu}$, $N_{uuud}$, $N_{uudd}$, $N_{uddd}$, $N_{duuu}$, $N_{duud}$, $N_{dudd}$ and $N_{dddd}$ are given numbers. So, a spins pair (1,1) appears with probability $2N_{uu}/(3N)$, which approximates to $(N_u/N)^2$. Using similar ideas, one can get a set of correlation functions of spins pair as follows

$$N_{uu} \approx 1.5N_u^2/N$$

$$N_{ud} \approx 3N_u(1-N_u/N)$$

$$N_{dd} \approx 1.5N(1-N_u/N)^2$$

$$N_{uuuu} \approx N_u^4/N^3$$

$$N_{uuud} \approx 3(1-N_u/N)N_u^3N^{-2}$$

$$N_{uudd} \approx 3N_u^2(1-N_u/N)^2N^{-1}$$

$$N_{uddd} \approx N_u(1-N_u/N)^3$$

$$N_{duuu} \approx N_u^3(1-N_u/N)N^{-2}$$

$$N_{duud} \approx 3N_u^2(1-N_u/N)^2N^{-1}$$

$$N_{dudd} \approx 3N_u(1-N_u/N)^3$$

$$N_{dddd} \approx (1-N_u/N)^4N \tag{2}$$

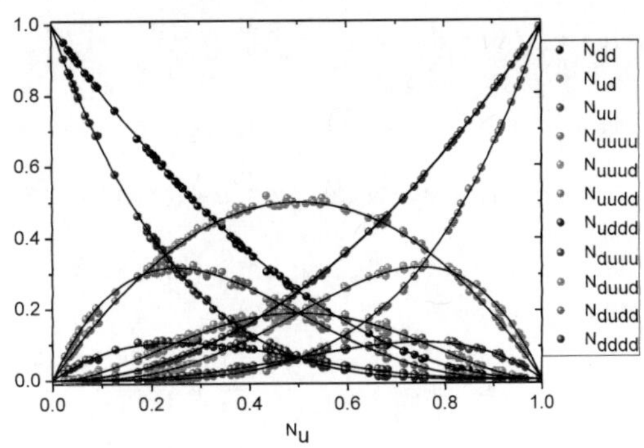


**Figure 2.** Various correlation functions of spins. The points are obtained from MC simulations for a randomly arranged lattice system on honeycomb with size $N=48^2$ and the solid lines are corresponding to the theoretical calculations. Every group of data is normalized by the total number of itself. There exists one solid line for the same two formulas in the right hand side of Eq. (2).

IOP Conf. Series: Materials Science and Engineering **378** (2018) 012004 doi:10.1088/1757-899X/378/1/012004

Meanwhile the Monte Carlo (MC) simulations for the spin correlation functions are also carried out, and the MC simulation and theory results based on Eq. (2) are shown in Fig. 2. It is easy to see that the theoretically calculated results are in very good accordance with the MC numerical simulation results. And these results are independent of the size of the lattice system, which indicates that even in the thermodynamic limit these correlation functions work very well.

Noticing that

$$\sum_{i=1}^{N} s_i = N_u - N_d = 2N_u - N$$

$$\sum_{<ij>} s_i s_j = N_{uu} + N_{dd} - N_{ud}$$

$$\sum_{<ijkl>} s_i s_j s_k s_l = N_{uuuu} - N_{uuud} + N_{uudd} - N_{uddd} - N_{duuu} + N_{duud} - N_{dudd} + N_{dddd} \quad (3)$$

and substituting Eq. (2) into Eq. (3), it becomes

$$\sum_{<ij>} s_i s_j \approx 3(N - 2N_u)^2 / 2N$$

$$\sum_{<ijkl>} s_i s_j s_k s_l \approx (N - 2N_u)^4 / N^3 \quad (4)$$

Let $L=2N_u/N-1$, using Eqs. (2) and (4), the Hamiltonian (1) can be rewritten as

$$H = (-1.5J_2L^2 - J_4L^4)N \quad (5)$$

Hence, the partition function of the system becomes

$$Q(T) = \sum_{\{s_i\}} \exp[\beta N(1.5J_2L^2 + J_4L^4)] \quad (6)$$

where $\beta = 1/kT$ and the sum is over all configurations. Noticing that the number of ways to choose $N_u$ out of $N$ is $N!/N_u!(N-N_u)!$, the partition function can be rewritten as

$$Q(T) = \sum_{L=-1}^{1} \frac{N!}{[N(1+L)/2]![N(1-L)/2]!} \exp[\beta N(1.5J_2L^2 + J_4L^4)] \quad (7)$$

In order to obtain the thermodynamic functions, Helmholtz free energy should be considered and calculated, i.e., $-kT\log_e(Q(T))$. Firstly, by taking the limit approximation, the $\log_e(Q(T))$ is replaced by the logarithm of the largest term in the summand when $N$ goes to infinity. Using Stirling's approximation, one can find that

$$\frac{1}{N}\log_e Q(T) \approx \beta(1.5J_2\bar{L}^2 + J_4\bar{L}^4) - \frac{1+\bar{L}}{2}\log_e\frac{1+\bar{L}}{2} - \frac{1-\bar{L}}{2}\log_e\frac{1-\bar{L}}{2} \quad (8)$$

where $\bar{L}$ is the value of $L$ that maximizes the summand of Eq. (7) and $\bar{L}$ is the root of the following equation

$$\log_e\frac{1+\bar{L}}{1-\bar{L}} = 2\beta(3J_2\bar{L} + 4J_4\bar{L}^3) \quad (9)$$

Thus one can deduce the phase transition properties of the system from the solution of Eq. (9). Series expansion technique and numerical methods are employed to solve it.

## 3. Numerical results and discussion

The main features of the solutions are illustrated in this section. Some special results will be presented here.

First, consider a simple situation, $J_2>0$, $J_4=0$, by treating Eq. (9) with series expansion technique, one can get critical temperature given by

$$kT_c = 3J_2 \tag{10}$$

such that

$$\bar{L} = \begin{cases} 0 & (T > T_c \\ \pm L_0 & (T < T_c) \end{cases} \tag{11}$$

Here, the results are the same as those of the Bragg-Williams approximation. The phase transition of the system is always continuous whatever the value $J_2$ taken.

Second, in the case of $J_2>0$, $J_4>0$, the things are very different from those stated in the first case. For small value of $J_4$, The phase transition is always continuous as in before case. However, a discontinuous transition is found for sufficiently high values of $J_4$. The crossover happens at value $J_4 = J_2/4$.

In the latter case, one can search carefully the roots of Eq. (9) by numerical method and series expansion technique. It will be found that its roots go to zero continuously if the temperature increases to the critical point in the case of $J_4 < J_2/4$. The phase transition is continuous. If they jump to zero, one has a discontinuous phase transition. In the situation of $J_4$ less than $J_2/4$, critical temperature is the same as indicated in Eq. (10). The critical exponent is the same as that obtained by traditional mean field method. In the situation of $J_4$ bigger than $J_2/4$ , the lower critical temperature is the same as Eq. (10) but the upper critical temperature $T_{uc}$ is given by the transcendental Eq. (12).

$$\begin{cases} tanh(\dfrac{3J_2\bar{L} + 4J_4\bar{L}^3}{kT_{uc}}) = \bar{L} \\ \dfrac{3J_2 + 12J_4\bar{L}^2}{kT_{uc}} = \dfrac{1}{1-\bar{L}^2} \end{cases} \tag{12}$$

## 4. Conclusions

In summary, the properties of the phase transition for Ising model on two dimensional honeycomb lattice system have been investigated. And the correlation functions of two-spin pairs and four-spin have been obtained, theoretical results are in good accordance with numerical simulation results.

The results have shown that the strength of four-spin interaction has deep affection on transition properties. It exhibits a crossover behaviour, namely when $J_4$ becomes greater than $J_2/4$, the phase transition is discontinuous.

It is believed that the theory presented in this work can also be applied to more complex geometrical situations, such as face-centred cubic lattice, body-centred cubic lattice, and so on. It can also be applied to more complex practical systems, such as Ising model in a uniform magnetic field, and can provide quantitative and qualitative results.

## 5. References

[1] N. De La Espriella, A.J. Arenas, M.S. Páez Meza 2016 *J. Magn. Magn. Mater.* **417** 434-441.
[2] B. Deviren, O. Canko, M. Keskin 2008 *J. Magn. Magn. Mater.* **320(18)** 2291-2299.
[3] N. Şarlı, S. Akbudak, Y. Polat, M.R. Ellialtıoğlu 2015 *Physica* A. **434** 194-200.
[4] U. Yu 2015 *Physica* A. **419** 75-79.
[5] D. F. de Albuquerque, I.P. Fittipaldi, J.R. de Sousa, N.O. Moreno, 2006 *Physica* B. **384**(1) 230-232.
[6] C. S. Prajapati, D. Visser, S. Anand, N. Bhat 2017 *Sens. Actuators,* B **252** 764-772.
[7] X. Shi, Y. Qi 2015 *J. Magn. Magn. Mater.* **393** 204-208.
[8] M. Wierzbicki 2017 *Physica* E **87** 220-227.
[9] Ü. Akıncı and G. Karakoyun 2017 *Physica* B. **521** 365-370.
[10] N. De La Espriella, C.A. Mercado, G.M. Buendía 2016 *J. Magn. Magn. Mater.* **417** 30-36.
[11] B. Teng, Y. Chen, H. Fu, Y. Tang, M. Tu, Y. Chen, J. Tang, 2002 *Solid State Commun.* **124(9)** 347-351.
[12] J. P. Santos, F.C. Sá Barreto 2017 *J. Magn. Magn. Mater.* **439** 114-119.
[13] Y. Yang, B. Teng, H. Yang, H. Cui 2017 *Physica.* A **483** 243-249.
[14] B. H. Teng, H.K. Sy 2006 *Europhys. Lett.* **73(4)** 601-606.
[15] Y. Yüksel, Ü. Akıncı, H. Polat 2012 *Physica* A. **391(3)** 415-422.
[16] Q. Zhang, G. Wei, Z. Xin, Y. Liang 2004 *J. Magn. Magn. Mater.* **280(1)** 14-22.
[17] B. Teng, H.K. Sy, 2004 *Solid State Commun.* **130(3)** 193-197.

IOP Conf. Series: Materials Science and Engineering **378** (2018) 012005 doi:10.1088/1757-899X/378/1/012005

# Synthesis of biocompatible carbon dots from yogurt and gas vapor sensing

**S Moonrinta[1], S Jamnongsong[2], S Sampattavanich[2], S Kladsomboon[3], W Sajomsang[4] and P Paoprasert[1*]**
[1]Department of Chemistry, Faculty of Science and Technology, Thammasat University, Pathumthani, 12120 Thailand
[2]Department of Pharmacology, Siriraj Hospital, Mahidol University, 2 Prannok Road, Bangkoknoi, Bangkok, 10700 Thailand
[3]Center for Research and Innovation, Faculty of Medical Technology, Mahidol University, Nakhon Pathom, 73170 Thailand
[4]National Nanotechnology Center, Pathumthani, 12120 Thailand
*Corresponding author, email: peerasak@tu.ac.th

**Abstract.** Carbon dots are fascinating nanomaterials due to their low cost, straightforward preparation, unique optical properties, and biocompatibility. In this work, carbon dots were prepared from yogurt using pyrolysis method. The carbon dots were characterized using UV-visible spectroscopy, fluorescence spectroscopy, Fourier transform infrared spectroscopy, and transmission electron microscopy. The carbon dots were applied for detecting formic acid vapor using electronic nose system. Furthermore, the biocompatibility of carbon dots were investigated to show their potentials for biomedical applications. Based on these research, carbon dots from yogurt were multifunctional fluorescent nanomaterials for various applications.

## 1. Introduction

In 2004, Xu and coworkers isolated a novel spherical carbon-based nanomaterial, from an eletrophoretic method for the purification of single-walled carbon nanotubes derived from arc-discharge soot.[1] This new material is now known as "carbon dots". Carbon dots are luminescent nanoparticles that have attracted broad attention in recent years. Carbon dots can fluoresce when they are excited by ultraviolet or visible light. In addition, they have low toxicity,[2] excellent photostability,[3] high water solubility[4] and can be prepared by several methods such as, laser ablation,[5] pyrolysis,[6] hydrothermal,[7] and microwave irradiation.[8] Because of these unique and attractive properties, carbon dots have been used in various applications such as sensing,[9] electronics,[10] and biomedical applications[7]. Carbon-dots have been synthesized from various precursors, such as wastes, edibles, and chemicals such as watermelon peel,[11] garlic[12] and ethanol.[13]

In this work, a method for synthesizing carbon-dots form yogurt was developed. As one of the oldest and most popular fermented foods, yogurt is a known delicacy around the world. It contains vitamins and minerals. It is also rich in protein, a source of nitrogen atoms known to enhance quantum efficiency of carbon dots. The carbon dots were synthesized from fermented milk via pyrolysis method. They were then characterized using UV-visible spectroscopy, fluorescence spectroscopy, Fourier transform infrared spectroscopy, and transmission electron microscopy. They were used in many sensing applications such as gas vapor detection using electronic nose system. In addition, they were also tested cytotoxicity with human bile duct cancer (KKU213) cell line.

## 2. Experimental section

### *2.1. Materials*
"Dutchie" original flavor yogurt was purchased from a local supermarket. All chemicals were purchased from Sigma-Aldrich. Cellulose dialysis membrane (1000 Da MWCO) was bought from Spectrum Labs. Deionized (DI) water was used throughout the experiment.

### *2.2. Methods and preparations*

#### *2.2.1. Synthesis of carbon dots*
Carbon dots were synthesized by pyrolysis method using yogurt as a precursor. Yogurt (50 g) and hydrochloric acid (0.5 M, 25 mL) were mixed in a 200 mL evaporating dish. The reaction mixture was heated at 180 °C for 2.5 h. Then remaining black solid was rinsed with DI water and adjusted pH 7. Next, the brown solution was purified using dialysis membrane for 3 days. The solution was then centrifuged for 15 minutes at 10000 rpm. After that the dot solution was dried by a freeze dryer overnight to give a brown solid and then kept in the bottle.

#### *2.2.2. Quantum yield*
Quantum yield (Φ) of carbon dots was measured using fluorescence spectrometry. Quinine sulfate was used as a reference. The carbon dots were dissolved in DI water (n = 1.33). Quinine sulfate was dissolved in 0.1 M $H_2SO_4$ solution (quantum yield of 54%, n = 1.33). Quantum yield was calculated using equation (1):

$$\Phi = \Phi_R \times \frac{I}{I_R} \times \frac{A_R}{A} \times \frac{n^2}{n_R^2} \quad (1)$$

where I is the integrated emission intensity, A refers to the absorbance, and n is refractive index of the solvent. The subscript R denotes the reference sample.

#### *2.2.3. Electronic nose measurement*
The optical electronic nose system consists of light sources, sensing materials, and a photo detector. Light-emitting diodes (LEDs) and a commercial photo detector (ET-TCS230) were used to detect light intensity. The eight gas sensor arrays were generated based on eight colors of LEDs. The transmitting light intensity through sensing thin film was observed in the form of photon frequency (Hz), captured by a photo-detector.

#### *2.2.4. Cytotoxicity test*
Human bile duct cancer (KKU213) cells, containing the nuclear marker NLS (nuclear localization signal)-mCherry, were chosen for the cell viability and imaging experiments. First, cells were placed in the wells of a microtiter plate at a ratio of 8,000 cells per well and incubated for 24 h at 37 °C under 5% $CO_2$. The media (150 μL per well) for cell culture was then replaced. Carbon dots (2.0 mg) were dissolved in a sterile 1X PBS solution (1 mL). The carbon dot solution was passed through a syringe filter (0.45 μm) and sonicated for 30 min at room temperature. The carbon dot solution (2 g $L^{-1}$, 100 μL) was then added to the first row of wells, and then diluted two-fold at each of the sequence of eight rows, to obtain carbon dots concentrations of 2.0000, 1.0000, 0.5000, 0.2500, 0.1250, 0.0625, 0.0313, 0.0156, and 0.0078 g $L^{-1}$. The last row of wells contained only cells and culture media with no carbon dot solution and was used as a control. The cells were then incubated for 24 h at 37 °C under 5% $CO_2$, which the culture media and carbon dot solution were removed. The incubated cells were washed three times with 1X PBS (200 μL). The incubated cells were then imaged (Operetta, 60X high NA objective

lens) using a 475 nm excitation wavelength and a 525 nm emission wavelength. The graph of relative viability against carbon dots concentration (0-2 g $L^{-1}$) was plotted from cell counting.

## 3. Result and discussion

### *3.1. Preparation and characterization of the carbon dots*

The carbon dots were prepared from yogurt by pyrolysis reaction. The functional groups of carbon dots were identified using FTIR. The FT-IR spectra indicated the characteristic absorption bands of O–H at 3278 $cm^{-1}$ and N–H stretching vibrations of amine groups at 2966 $cm^{-1}$ and C=O stretching vibrations located at around 1640 $cm^{-1}$ (Figure 1a). This supports that carbon dots are highly hydrophilic and water soluble. Size and morphology of carbon dots were determined by TEM. The TEM image showed spherical carbon dots with a diameter of about 3.5 nm (Figure 1b). The optical properties of carbon dots were characterized using UV-visible and fluorescence spectroscopy. The UV–visible spectrum exhibited a broad absorption band from 250 to 500 nm (Figure 1c). The absorption peak centered at 282 and 359 nm, corresponding to $\pi$-$\pi$* transition of the aromatic $sp^2$ domains and n-$\pi$* transition of C=O bonds due to the presence of carbonyl-based groups on the surface.[14] The emission was found to be excitation wavelength dependence, showing the highest emission intensity at 320 nm excitation wavelength (Figure 1d). The carbon dots showed blue emission under UV illumination with a quantum yield of 2.4%.

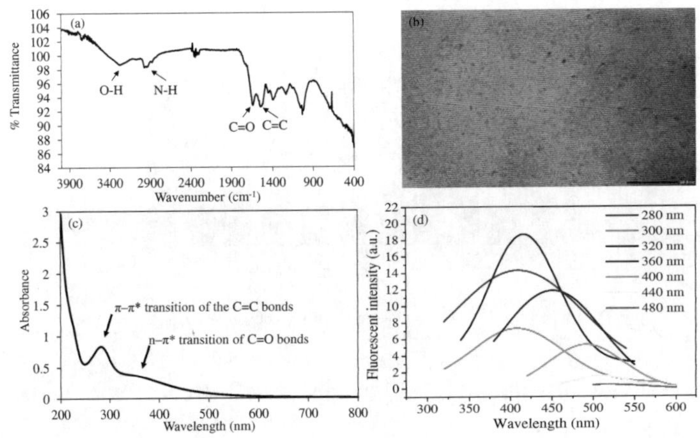


**Figure 1.** (a) FT-IR spectrum, (b) TEM image, (c) absorption, and (d) excitation-dependent emission spectra of carbon dots.

### *3.2. Sensing applications of the carbon dots*

#### *3.2.1. Formic acid vapor sensing*

Carbon dots were used to detect chemical vapor, in this case, formic acid. The dynamic optical responses of the film (Figure 2a and b) corresponded to a change in the light intensity. When the nitrogen reference gas flowed through the film, the light transmission in the frequency unit increased rapidly. The light transmission then dropped dramatically when the flow was switched to reference gas. The carbon dots showed different responses when 10% and 20% formic acid solutions were used. This suggested that the carbon dots were successfully tested as a formic acid vapor sensor.

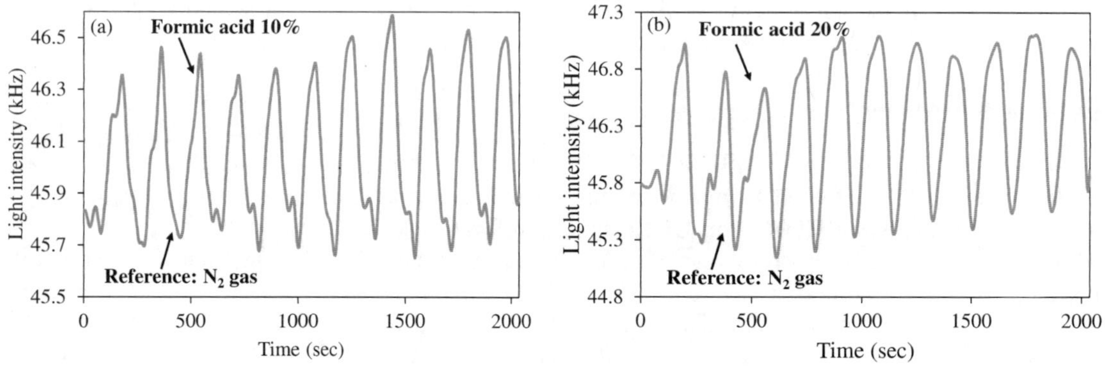


**Figure 2.** Gas sensing responses with carbon dots film for (a) formic acid 10%, and formic acid 20%.

*3.2.2. Cytotoxicity*

In this work, KKU213 cell line, a human bile duct cancer, was used to test the cytotoxicity of carbon dots. Cells were treated with different concentrations of the carbon dots for 24 h. The relative cell viability showed more than 90% up on 2 g $L^{-1}$ concentration of carbon dots (Figure 3). The result suggested that carbon dots showed excellent biocompatibility.

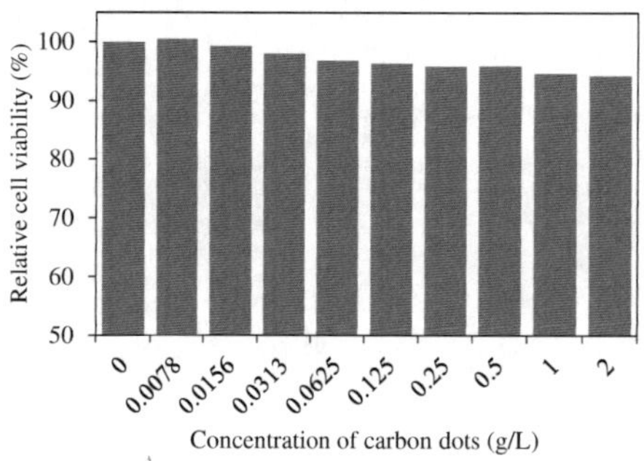


**Figure 3.** Relative cell viability human bile duct cancer cells treated with carbon dots for 24 h.

## 4. Conclusion

From this study, carbon dots were prepared from yogurt using a simple pyrolysis method. The carbon dots have been successfully applied for formic acid vapor. In addition, they exhibited excellent biocompatibility in living cell. The multifunctional carbon dots from yogurt developed in this work will be suitable for sensing and biomedical applications. Future work includes cell imaging and chemical sensing using our carbon dots.

## 5. Acknowledgement

This research is supported by the governmental budget through Thammasat University in the fiscal year 2017 (Contract No. 6/2560). S. Moonrinta is sponsored by the Thailand Graduate Institute of Science and Technology (TGIST: SCA-CO-2559-2322-TH), National Science and Technology Development Agency (NSTDA). The authors acknowledge the Central Scientific Instrument Center (CSIC), Division of Chemistry, Division of Biotechnology, Faculty of Science and Technology, and Thammasat University.

## 6. References

[1] Xu, X., et al., *J Am Chem Soc*, 2004. **126**(40): p. 12736-7.
[2] Yang, S.T., et al., *J Phys Chem C Nanomater Interfaces*, 2009. **113**(42): p. 18110-4.

[3] Baker, S.N. and G.A. Baker, *Angew Chem Int Ed Engl*, 2010. **49**(38): p. 6726-44.
[4] Qu, S., et al., *Angew Chem Int Ed Engl*, 2012. **51**(49): p. 12215-8.
[5] Goncalves, H., et al., *Sensors and Actuators B-Chemical*, 2010. **145**(2): p. 702-7.
[6] Zhu, L.L., et al., *Journal of Materials Chemistry C*, 2013. **1**(32): p. 4925-32.
[7] Sahu, S., et al., *Chem Commun (Camb)*, 2012. **48**(70): p. 8835-7.
[8] Zhai, X., et al., *Chem Commun (Camb)*, 2012. **48**(64): p. 7955-7.
[9] Yan, F., et al., *Sensors and Actuators B: Chemical*, 2014. **192**: p. 488-95.
[10] Han, X., et al., *Dalton Trans*, 2013. **42**(29): p. 10380-3.
[11] Zhou, J.J., et al., *Materials Letters*, 2012. **66**(1): p. 222-4.
[12] Sun, C., et al., *Nanoscale Res Lett*, 2016. **11**(1): p. 110.
[13] Li, H.T., et al., *New Journal of Chemistry*, 2011. **35**(11): p. 2666-70.
[14] Wang, Y. and A. Hu, *Journal of Materials Chemistry C*, 2014. **2**(34): p. 6921-39.

IOP Conf. Series: Materials Science and Engineering **378** (2018) 012006 doi:10.1088/1757-899X/378/1/012006

# Imperfections in carbon nanotubes structure and their impact on the basic mechanical properties

**Marino Brčić**[1, 2], **Marko Čanađija**[2] **and Josip Brnić**[2]
[1]Corresponding author: mbrcic@riteh.hr
[2]University of Rijeka, Faculty of Engineering, Vukovarska 58, HR-51000 Rijeka, Croatia

**Abstract.** Theoretical and experimental research of nanocomposite materials have shown that usage of carbon nanotubes as a reinforcement, significantly improves the mechanical properties of aforementioned composites. Carbon nanotubes rarely appear in ideal form. Different defects within nanotube structure such as vacancy defects, or waved shape of the nanotube, can greatly influence the mechanical properties of carbon nanotube and thus, decrease the final mechanical properties of carbon nanotube reinforced composites. The paper at hand investigates degradation of basic mechanical properties of single and double walled carbon nanotubes, straight and waved, with different vacancy and topological defects. Also, various nanotube patterns are considered.

## 1. Introduction

Very soon after the landmark paper of Iijima in 1991 [1], carbon nanotubes (CNTs) attracted considerable research attention in many theoretical and experimental researches, with main goal to find out their properties and characteristics. Common conclusion, so far, is that CNTs possess extraordinary mechanical, thermal and electrical properties. Those facts make them a logical choice as a reinforcement in composite material, either to improve mechanical properties of composites [2], or to improve electrical conductivity [3]. Due to their size, computer modelling and numerical methods are logical solution for investigating the mechanical properties of CNTs and nanocomposite materials. One of the numerical approaches is based on structural mechanics, where covalent bonds between atoms in nanotube structure are replaced with finite elements and nodes [4, 5, 6]. Initially, research on CNTs was directed toward determination of mechanical properties of ideally shaped CNTs [7, 8, 9], but more careful research detected that single walled (SWNT) and multiwalled (MWNT) carbon nanotubes rarely come in ideal form and without defects. Regardless of the manner of production control and methods, CNTs are usually waved [10, 11], and studies have shown that the waviness of the nanotube has an influence on the final mechanical properties of CNTs and, thus, nanocomposite materials [12, 13, 14]. Another common problem in nanotube structure is appearance of various defects [15, 16, 17], which can occur either naturally or artificially. All of the aforementioned leads to the necessity for detailed study of the impact of defects and waviness of CNTs on their mechanical properties, specifically on basic mechanical properties like elastic modulus, since it eventually leads to the decrease of final mechanical properties of the nanocomposites. The paper at hand investigates the change of longitudinal elastic moduli of SWNT and DWNT, both straight and waved, with different defects. Different pattern of CNTs are considered, i.e. armchair, zig-zag and combined chiral pattern.

IOP Conf. Series: Materials Science and Engineering **378** (2018) 012006 doi:10.1088/1757-899X/378/1/012006

## 2. Imperfections (waviness, vacancies and defects)

CNTs can be classified as SWNT, double (DWNT) or MWNTs. Their aspect ratio (diameter/length ratio) is very high, with diameter in order of nanometers and length in micrometers. High aspect ratio leads to waved shape of CNTs, which is experimentally proved [15, 16], and to small bending stiffness. Waviness of CNTs is in literature defined with the waviness ratio $w = a / l$ [18], where a represents wave amplitude and l wavelength. Mechanical properties of CNTs are influenced not only by waviness but also by various defects within nanotube structure. Those defects can occur either naturally or artificially. Mainly, defects can be classified in four different groups [10, 19]: topological, rehybridization, incomplete bonding defects and doping with other atoms than carbon. Topological defects imply that instead of hexagons in CNT structure, other mesh form can occur (heptagons, pentagons and similar). Focus in this paper will be on topological and incomplete bonding defects (vacancies or "missing" atoms). Research, experimental and theoretical [10, 17], have shown that topological Stone-Wales (SW) or 5-7-7-5 defects, are commonly present in CNTs and randomly distributed, thus altering the elastic properties of CNTs. SW defects are composed from two pentagons and two heptagons. This defect is formed by rotating one carbon bond by 90°. Defects classified as incomplete bonding defects, vacancies, can occur in CNTs structure for example during a purification process or by irradiation.

## 3. Nanotube examples

Several different CNT models, single and double walled, have been prepared in this paper. Influence of imperfections is shown on FEM model of following nanotube models: SWNT, armchair (5, 5) and zig-zag (9, 0) pattern; DWNT, armchair (4,4/9,9), zig-zag (7,0/16,0), chiral A (4,4/16,0), chiral B (7,0/9,9) patterns. Chiral A pattern denotes a DWNT with inner armchair tube and outer zig-zag, and accordingly, chiral B denotes a DWNT with inner zig-zag tube, and outer armchair. The selected SWNTs have similar diameter (0,678nm – 0,695nm) and aspect ratio (20,2). The same goes for DWNTs, similar inner (0,535nm – 0,54nm) and outer (1,20nm – 1,23nm) diameter and aspect ratio (20,1). All CNTs are modelled as a space frame structure, where atoms are replaced by nodes, and covalent bonds with the beam finite elements [4, 5, 6]. In DWNTs, van der Waals (VDW) interactions are formed between atoms in inner and outer tube. Those VDW interactions between two different DWNT layers were modelled with rod finite elements, with properties adjusted to match the forces which originate from Lennard – Jones potential. To determine the elastic modulus, an axially loaded nanotube model was used, with 1 nN axial force used in all examples. The longitudinal elastic modulus was obtained using classical term $E = (F \cdot l / A \cdot \Delta l)$, where l represents nanotube length and $\Delta l$ the elongation obtained from FEM analysis. The number of finite elements rises from around 2350 elements in single SWNT model to 355000 elements in DWNT model, due to the VDW interactions. Five models have been prepared for each SWNT pattern, with different waviness ratio, i.e. four waved and one straight, and three models for each DWNT pattern (one straight and two waved), as shown in figure 1.

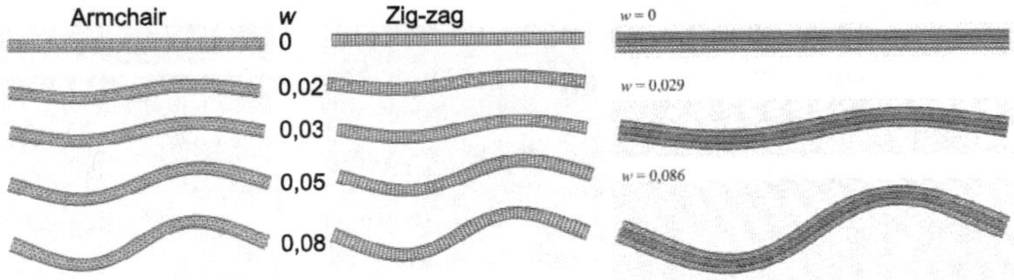


a) b)

**Figure 1.** Examples of waved SWNTs (a) and DWNTs (b).

In the first defect case studied in this paper, randomly selected percentage of missing atoms (0,1%, 0,5%, 1%, 2%, 5%, 10%) were removed from CNTs structure. Vacancies are obtained by deleting one node and corresponding three beam elements from CNT structure. That was done for both, SWNTs and DWNTs, for each shape. In the second studied case, 5-7-7-5 (SW) defects were implemented into CNTs structure, with 2, 4, 6, 8 and 10 of those defects in each SWNT and the same amount of SW defects in DWNT, but in inner and outer layer altogether. The second case was implemented on armchair and zig-zag SWNT, with waviness ratio 0,05, and on all patterns of DWNT, with waviness ratio 0,029.

## 4. Results

Results for longitudinal elastic modulus obtained from aforementioned examples are given in following tables and figures. Tables 1. to 3. show the results for the first vacancy case, for armchair and zig-zag SWNT, armchair and zig-zag DWNT, chiral A and chiral B DWNT, respectively. As expected, there is a noticeable drop of longitudinal elastic modulus with the increase of the waviness ratio of the nanotube, regardless of the CNT type and pattern. The value of elastic modulus also decreases with the increase of the vacancies within the same waviness ratio of the CNT, but is practically unaffected when percentage of defects is less than 1%. Results for elastic modulus of defect free SWNT, both patterns, table 1, as well as for DWNT, table 2, coincide well with the results given in the literature [20, 21, 22]. When it comes to SW defects, a decrease of elastic modulus values can be noticed for both, SNWTs and DWNTs, but that decrease is not as drastic as in vacancy case, figure 2.

**Table 1**. Results for first vacancy case, armchair and zig-zag SWNT

| | Longitudinal elastic modulus *E*, GPa | | | | | | | | | |
|---|---|---|---|---|---|---|---|---|---|---|
| | Waviness ratio, *w* (armchair): | | | | | Waviness ratio, *w* (zig-zag): | | | | |
| Vacancy %: | 0 | 0,02 | 0,03 | 0,05 | 0,08 | 0 | 0,02 | 0,03 | 0,05 | 0,08 |
| No defects | 1048,76 | 526,22 | 341,58 | 135,09 | 74,27 | 1030,99 | 533,16 | 349,11 | 139,97 | 77,45 |
| 0,1% | 1008,82 | 526,38 | 335,58 | 132,74 | 72,96 | 987,36 | 533,14 | 342,38 | 138,77 | 73,05 |
| 0,5% | 979,79 | 511,43 | 316,59 | 127,16 | 71,98 | 918,42 | 523,70 | 328,33 | 129,49 | 74,78 |
| 1% | 667,74 | 456,06 | 313,61 | 126,50 | 69,89 | 848,07 | 516,36 | 310,91 | 124,27 | 73,59 |
| 2% | 796,56 | 434,06 | 285,31 | 117,67 | 60,25 | 818,34 | 421,74 | 299,12 | 97,38 | 56,66 |
| 5% | 662,10 | 331,67 | 219,99 | 76,94 | 47,99 | 446,47 | 246,32 | 87,64 | 103,91 | 47,93 |
| 10% | 146,87 | 57,89 | 74,42 | 59,98 | 13,91 | 282,99 | 88,98 | 60,26 | 36,85 | 18,77 |

**Table 2.** Results for first vacancy case, armchair and zig-zag DWNT

| | Longitudinal elastic modulus *E*, GPa | | | | | |
|---|---|---|---|---|---|---|
| | Waviness ratio, *w* (armchair): | | | Waviness ratio, *w* (zig-zag): | | |
| Vacancy %: | 0 | 0,029 | 0,086 | 0 | 0,029 | 0,086 |
| No defects | 1111,82 | 355,06 | 55,57 | 1104,17 | 381,76 | 58,93 |
| 0,1% | 1098,17 | 348,81 | 54,70 | 1080,36 | 373,29 | 58,75 |
| 0,5% | 1067,27 | 344,32 | 53,43 | 1047,19 | 374,03 | 57,57 |
| 1% | 982,84 | 335,69 | 49,47 | 1007,07 | 381,49 | 51,64 |
| 2% | 856,31 | 314,64 | 47,48 | 886,28 | 341,62 | 47,95 |
| 5% | 599,24 | 209,10 | 33,94 | 676,09 | 259,06 | 40,57 |
| 10% | 404,66 | 140,67 | 22,98 | 377,76 | 185,33 | 21,15 |

**Table 3.** Results for first vacancy case, Chiral A and Chiral B DWNT

| | Longitudinal elastic modulus *E*, GPa | | | | | |
|---|---|---|---|---|---|---|
| | Waviness ratio, *w* (Chiral A): | | | Waviness ratio, *w* (Chiral B): | | |
| Vacancy %: | 0 | 0,029 | 0,086 | 0 | 0,029 | 0,086 |
| No defects | 1150,27 | 362,87 | 60,69 | 1089,91 | 368,53 | 55,90 |
| 0,1% | 1100,68 | 360,06 | 60,54 | 1078,54 | 364,33 | 55,57 |
| 0,5% | 1065,38 | 351,03 | 58,46 | 1041,90 | 358,36 | 50,65 |
| 1% | 1059,48 | 325,77 | 57,03 | 982,24 | 347,49 | 53,03 |
| 2% | 893,37 | 279,15 | 56,43 | 903,22 | 307,77 | 49,00 |
| 5% | 667,07 | 257,85 | 45,19 | 635,40 | 282,38 | 36,96 |
| 10% | 393,66 | 156,01 | 22,96 | 382,70 | 168,95 | 15,94 |

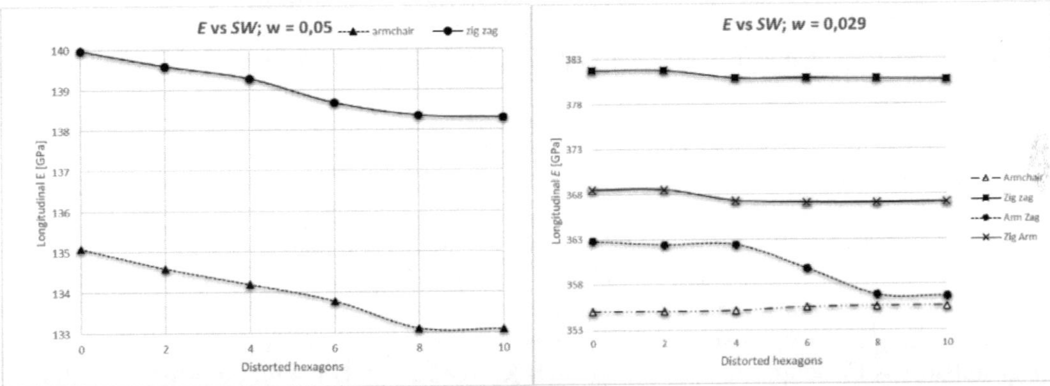


a) b)

**Figure 2.** The dependence of longitudinal modulus on SW defects, SWNTs(a) and DWNTs (b).

## 5. Conclusions

The impact of the various imperfections within carbon nanotube structure on the basic mechanical properties of CNTs is shown and analysed in this paper, on FEM model of armchair and zig-zag SWNT and four patterns of DWNT. Also, this was done on CNTs with different waviness ratio. Final conclusion is that CNTs possess excellent mechanical properties, and are quality choice as a reinforcement in the nanocomposite materials, but as it is shown, various defects and geometrical characteristics can affect their properties, and thus decrease the final mechanical properties of nanocomposite. The effect of SW defects is lesser when compared to vacancies. Future research will be focused on theoretical modelling of nanocomposite materials, reinforced with SWNTs and DWNTs, based on FEM, to study the impact of nanotubes imperfections on nanocomposite mechanical properties. Also, the experimental research of nanocomposite materials is planned, to confirm the theoretical results and conclusions.

## 6. Acknowledgments

This work has been financially supported by the University of Rijeka under the project no. 16.09.2.1.03 and by Croatian Science Foundation under the project no. 6876. This support is gratefully acknowledged.

## 7. References

[1] Iijima S 1991 Nature 354 56-58

[2] Tserpes K I, Papanikos P, Labeas G, Pantelakis S G 2008 *Theoretical and Appliled Fracture Mechanics* 49 51-60
[3] Murarescu M, Dima D, Andrei G, Ciriumaru *A 2011 Annals of DAAAM for 2011 & Proceedings of 22nd International DAAAM Symposium* 22 925-6
[4] Brcic M, Canadija M, Brnic J 2009 *Estonian Journal of Engineering* 15 79-86
[5] Brcic M, Canadija M, Brnic J 2013 *Review on Advanced Materials Science* 33 1-6
[6] Brcic M, Canadija M, Brnic J 2013 *Meccanica48 2209-20*
[7] Fisher F T, Bradshaw R D, Brinson L C 2002 *Applied Physics Letters* 80 4647-49
[8] Fisher F T, Bradshaw R D, Brinson L C 2003 *Composites Science and Technology* 63 1689-1703
[9] Bradshaw R D, Fisher F T, Brinson L C 2003 *Composites Science and Technology* 63 1705-22
[10] Ebbesen T, Takda T 1995 *Carbon* 33 973-8
[11] Shokrieh M M, Rafiee R 2010 *Composite Structure* 92 647-52
[12] Jiang L Y, Huang Y, Jiang H, Ravichandran G, Gao H, Hwang K C 2006 *Journal of the Mechanics and Physics of Solids* 54 2436-52
[13] Tan H, Jiang L Y, Huang Y, Liu B, Hwang K C 2007 *Composite Science and Technology* 67 2941-46
[14] Qian D, Wagner G J, Liu W K, Yu M, Ruoff R S 2002 *Appl. Mech. Rev* 55 495-533
[15] Shaffer M S P, Windle A H 1999 *Adv. Mater.* 11 937-41
[16] Vigolo B, Penicaud A, Coulon C, Sauder C, Pailler R, Journet C, Bernier P, Poulin P 2000 *Science* 290 1331-34
[17] Patel A M, Joshi A Y 2016 *Procedia Technology* 23 122-9
[18] Farsadi M, Ochsner A, Rahmandoust M 2012 *Journal of Composite Materials* 47 1-10
[19] Charlier J 2002 *Accounts of Chemical Research* 35 1063-69
[20] Chandra N, Namilae S, Shet C 2004 *Physical Review* 69
[21] Kalamkarov A L, Georgiades A V, Rokkam S K, Veedu V P, Ghasemi-Nejhad M N 2006 *International Journal of Solids and Structures* 43 6832-54
[22] Song H Y, Zha X W 2008 *Journal of Computational and Theoretical Nanoscience* 5 1481-84

IOP Conf. Series: Materials Science and Engineering **378** (2018) 012007 doi:10.1088/1757-899X/378/1/012007

# Comparison of the mechanical behavior of materials subjected to specific operating conditions

**J Brnic[1,*], S Krscanski[1] and M Brcic[1]**

[1]Faculty of Engineering, University of Rijeka, Vukovarska 58, 51000 Rijeka, Croatia
[*]Author to whom any correspondence should be addressed

E-mail: brnic@riteh.hr

**Abstract.** The materials for making the construction elements are selected according to the required conditions and the available material properties. In this sense mechanical properties and material behavior of the materials such as 42CrMo4 steel, X46Cr13 steel and X6CrNiTi18-10 steel were investigated. Mechanical properties of the mentioned materials are shown in the form of engineering stress-strain diagrams, creep behavior is presented in the form of creep curves while uniaxial fatigue of these materials are given using life diagrams. These materials differ in chemical composition, their strength properties differ within 20% while their fatigue limits are significantly different.

## 1. Introduction

Design of the considered element must be optimal. This means that it is necessary to select a material corresponding to the required operating conditions. During the service life of the structure many failures can appear [1-3]. Some of them may arise as a consequence of operating conditions (structural loading, corrosion, etc.) and some of them may arise as assembly errors, improper maintenance, manufacturing defects, design errors, etc. In any case it is needed to determine why and how an engineering component has failed. Since creep and fatigue are possible failures, they are subject of this research. It is known that creep phenomenon is usually defined as time-dependent behavior where strain continuously increases while the stress (load) is kept constant [4]. Creep is usually appreciable at temperature above 40% of the melting temperature of the considered material [5]. In these investigations, some materials were subjected to creep at different stress levels and different temperatures. From the other hand, these materials were also subjected to uniaxial fatigue at stress ratio R = 0.25. More data about considered materials can be found in literature, and some of the published papers are mentioned here, [6-7, 8-9, 10-11]. Some data related to fatigue resistance of 42CrMo4 steel are presented in Ref. [12]. Further, Ref. [13] focuses on the lifetime reduction of cyclically loaded X46Cr13 steel constantly exposed to highly corrosive CO2-saturated hot thermal water at 60°C. The damage of the material and the microhardness variations of X6CrNiTi18-10 were analyzed at different cavitation conditions in a cavitation chamber, Ref. [14]. The main intention of this research was to obtain mechanical properties at different temperatures of these materials as well as consider their creep resistance and fatigue limit.

## 2. Considered materials, equipment, specimens, standards

Tested materials were, table 1: 42CrMo4 steel (chromium-molybdenum – manganese low-alloy constructional steel, 16 mm, soft annealed and cold drawn), applications: statically and dynamically stressed machine components (gears, crankshafts, etc.); X46Cr13 steel (chromium martensitic stainless steel, 16 mm annealed and cold drawn round steel), applications: in manufacturing of pump parts, machined metal parts, shaft, gears, tie rods, screws; X6CrNiTi18-10 steel (18 mm soft annealed austenitic stainless steel round bar, applications are as follows: automotive industry, chemical industry (chemical processing equipment), building and construction industries, food and beverage industries and aviation and aerospace industries.

**Table 1.** Tested materials: Chemical compositions (only some elements are shown)

| Material | Chemical composition: some of the elements | | | | | |
|---|---|---|---|---|---|---|
| | C | Cr | Si | Mn | Mo | Ni |
| 42CrMo4; **1.7225;** AISI 4140 | 0.42 | 1.07 | 0.24 | 0.84 | 0.22 | - |
| X46Cr13; **1.4034;** AISI 420 | 0.442 | 13.05 | 0.375 | 0.381 | 0.049 | - |
| X6CrNiTi18-10; **1.4541;** AISI 321 | 0.0176 | 16.95 | 0.436 | 1.44 | 0.241 | 9.236 |

Equipment used: materials testing machine (Zwick/Roell) of 400 kN, macro-extensometer, high temperature extensometer, furnace (900°C), dynamic testing machine (Servopulser), (±50kN).
Specimens for tensile testing related to stress-strain diagrams and tensile creep behavior were manufactured in accordance with ASTM Standard, ASTM: E 8M-15a, and for fatigue axial testing in accordance with ASTM: E 466-15 standard.
Standard used in tensile testing related to stress-strain procedure at room temperature was ASTM: E 8M-15a, while that used at high temperatures was ASTM: E21-09. Creep tests were carried out in accordance with ASTM: E 139-11 standard. Fatigue tensile testing has been carried out according to ISO 2017 standard. Also, all mentioned ASTM standards can be found in Annual Book of ASTM Standards (2015), [15].

## 3. Experimental results

*3.1 Determination of engineering stress-strain diagrams*
Engineering stress-strain diagrams were obtained on the basis of tensile tests which were performed at different temperatures, figure 1.

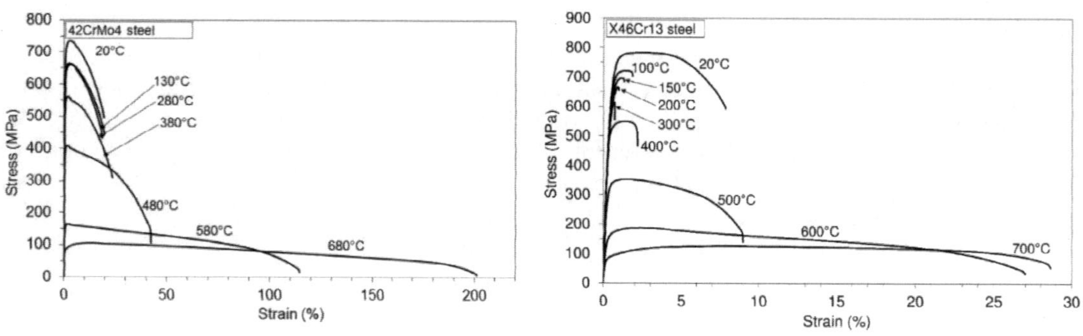

IOP Conf. Series: Materials Science and Engineering **378** (2018) 012007 doi:10.1088/1757-899X/378/1/012007

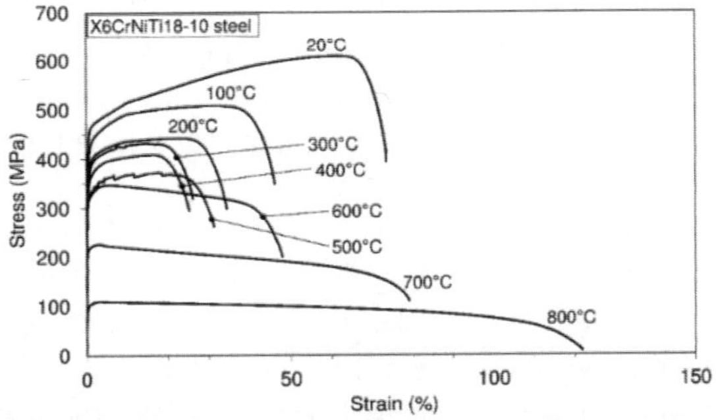


**Figure 1.** Engineering stress-strain diagrams at different temperatures: 42CrMo4, X46Cr13, X6CrNiTi18-10.

Based on these diagrams mechanical properties (ultimate tensile strength, yield strength, modulus of elasticity) as well as total elongation and contraction in area of the specimen of considered steels were determined. As it is visible, steel X46Cr13 has the highest ultimate tensile strength [6, 8, 10].

*3.2. Short-time creep tests*

Short-time creep tests show the behavior of the considered material during considered time period. This behavior of the material is very important property that defines the possibility of use of the material at higher temperature range. As it is known only 1-2% of creep strains are allowed in engineering practice. Consideration and knowledge of the creep behavior of material is of importance for material that can be exposed to high temperature in short time such as fire or other hazard environmental conditions. In figure 2 creep behavior of considered materials is shown.

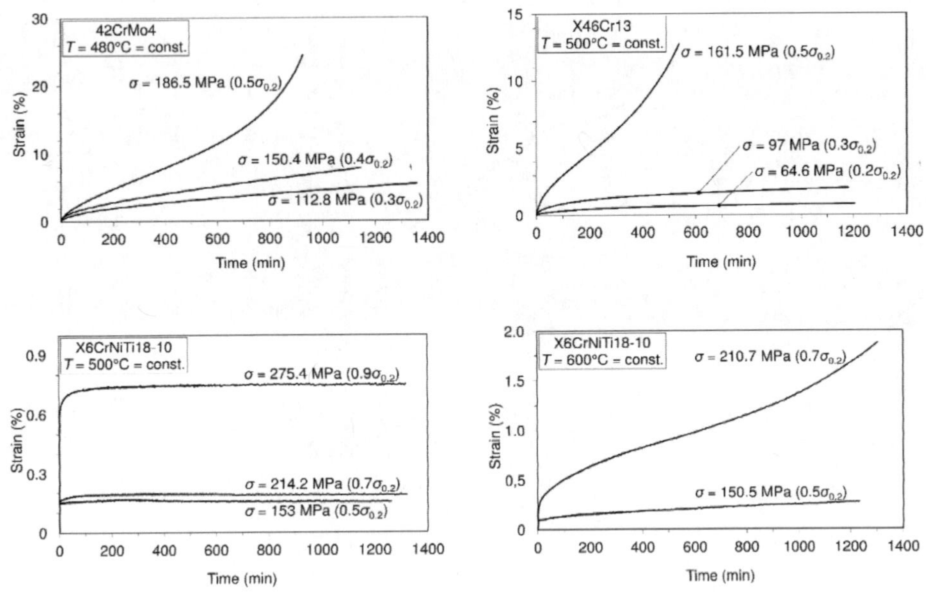


**Figure 2.** Creep behavior at different stress levels and temperatures: 42CrMo4, X46Cr13, X6CrNiTi18-10.

Creep resistance of considered materials can be assessed based on the performed short-time creep tests, [6, 8, 10]. Based on known creep curves sometimes is of interest to model creep behavior using analytical formulas or rheological models. These procedures are more useful since experimental investigations consume much time and money.

IOP Conf. Series: Materials Science and Engineering **378** (2018) 012007 doi:10.1088/1757-899X/378/1/012007

### *3.3 Fatigue testing*

Fracture of an engineering element is usually defined as the process of damage accumulation due to cyclic loading during its service life [16]. In this investigation uniaxial cyclic loading was considered. Tests were carried out at stress ratio R = 0.25 at all of examined materials and at R = – 1 at steel X46Cr13. Tests were performed on unnotched specimens in accordance with the ISO 12107:2012 (2012) standard, [17]. As the results of the fatigue tests, experimentally obtained data are placed in a coordinate system, figure 3. On the ordinate ($\sigma_{max}$/MPa) data related to the maximum stresses is plotted, while on the abscissa (N) data related to the number of the cycles to failure is plotted. In this way, so called S – N diagram (stress versus number of cycles to failure), or diagram known as Wohler curve or fatigue life diagram can be created, [6, 8, 10]. Fatigue limit (endurance limit) was calculated by modified staircase method.

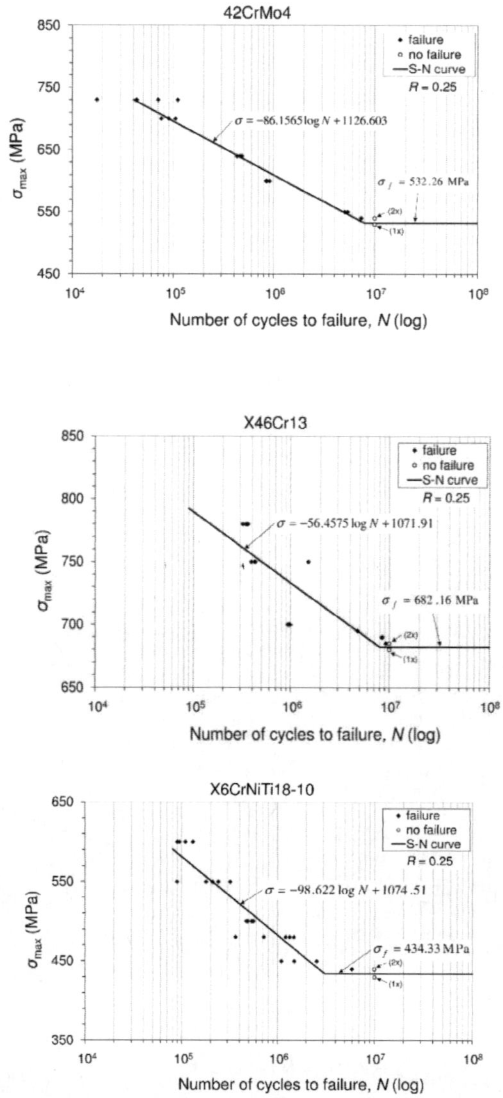


**Figure 3.** Stress versus number of the cycles to failure: *S – N* curves.

Based on the obtained results it is visible that the lowest fatigue limit has material X6CrNiTi18-10, when consideration is taken for stress ratio *R* = 0.25.

## 4. Conclusion

Mechanical properties, creep resistance and fatigue limit were determined for considered materials based on experimental investigations. These data can be of interest for designers who are involved in structural design using considered materials. Based on these investigations it is visible that material X6Cr13 has the highest ultimate tensile strength at room temperature, material X6CrNiTi18-10 may be treated as most creep resistant, while material X6Cr13 has the highest fatigue limit. Some other investigations related to Charpy impact energy, fracture toughness as well as fatigue limit at elevated temperature can be of interest.

## 5. Acknowledgments

This work has been supported in part by Croatian Science Foundation under the project 6876 and it is also partially supported by the University of Rijeka within a project 13.09.1.1.01.

## 6. References

[1] Brooks C R and Choudhury A 2002 *Failure Analysis of Engineering Materials* (USA: McGraw-Hill)

[2] Farahmand B, Bockrath G and Glassco J 1997 *Fatigue and Fracture Mechanics of High Risk Parts* (New York: Chapman & Hall)

[3] Collins J A 1993 *Failure of Materials in Mechanical Design 2nd edition* (New York: John Wiley & Sons)

[4] Findley W N, Lai J S and Onaran K 1989 *Creep and Relaxation of Nonlinear Viscoelastic Materials* (New York: Dover Publication Inc.)

[5] Raghavan V 2004 *Materials science and engineering* (New Delhi: Prentice- Hall of India)

[6] Brnic J, Canadija M, Turkalj G, Krscanski S, Lanc D, Brcic M and Zeng G 2016 Short-time creep, fatigue and mechanical properties of 42CrMo4–low alloy structural steel *Steel Compos. Struct.* 4 **22** 875–88.

[7] Terres M A, Ben Mohamed S and Sidhom H 2010 Influence of ion nitriding on fatigue strength of low-alloy (42CrMo4) steel: experimental characterization and predictive approach *Int. J. Fatigue* 11 **32** 1795–804.

[8] Brnic J, Krscanski S, Lanc D, Brcic M, Turkalj G, Canadija M and Niu J 2017 Analysis of the mechanical behavior, creep resistance and uniaxial fatigue strength of martensitic steel X46Cr13 *Materials* 4 **10** 388–406.

[9] Rosemann P, Kauss N, Muller C and Halle T 2015 Influence of solution annealing temperature and cooling medium on microstructure, hardness and corrosion resistance of martensitic stainless steel X46Cr13 *Mater. Corros.* 10 **66** 1068–76.

[10] Brnic J, Turkalj G, Canadija M, Lanc D, Krscanski S, Brcic M, Li Q and Niu J 2016 Mechanical properties, short time creep and fatigue of an austenitic steel *Materials* 4 **9** 298-1–298-19.

[11] Celik O, Baydogan M, Atar E, Kayali S and Cimenoglu H 2013 Fatigue performance of low temperature nitrided AISI 321 grade austenitic stainless steel, *Mater. Sci. Eng.* A. 565 **10** 38–43.

[12] Černy I, Sis J 2014 Fatigue Strength of Laser Hardened 42CrMo4 Steel Considering Effects of Compressive Residual Stresses on Short Crack Growth, *Procedia Engineering* **74** 417-420.

[13] Pfennig A, Wiegand R, Wolf M, Bork C P 2013 Corrosion and corrosion fatigue of AISI 420C (X46Cr13) at 60 degrees C in CO2-saturated artificial geothermal brine, *Corrosion Science* **68** 134-143.

[14] Krella A 2005 Influence of cavitation intensity on X6CrNiTi18-10 stainless steel performance in the incubation period, *Wear* 11–12 **258** 1723-1731.

[15] *Metallography: Metals–Mechanical Testing: Elevated and Low–Temperature Tests (Annual Book of ASTM Standards vol. 03.01)* 2015 (ASTM International: Baltimore)

[16] Pollak R D 2005 *Analysis of methods for determining high cycle fatigue strength of a material with investigation of Ti-6Al-4V gigacycle fatigue behavior*, dissertation (Wright-Patterson Air Force Base, Ohio: Air Force Institute of Technology)
[17] *12107:2012-Metallic Materials-Fatigue Testing-Statistical Planning and Analysis of Data* 2012 (Geneva: International Organization for Standardization)

IOP Conf. Series: Materials Science and Engineering **378** (2018) 012008 doi:10.1088/1757-899X/378/1/012008

# Mechanical Properties of Cf/SiC Composites with Ameliorated Interface Employing a New Precursor

**Y L SHI and K JIAN**

Science and Technology on Advanced Ceramic Fibers and Composites Laboratory, National University of Defense Technology, Changsha, China

E-mail: Shiyunliang_nudt@163.com

**Abstract.** Three dimensional and four directional carbon fiber reinforced SiC matrix (Cf/SiC) composites were fabricated by precursor infiltration and pyrolysis process with ameliorated interface employing a new precursor, liquid polyvinylcarbosilane (LPVCS). LPVCS is a novel precursor with active Si-H and –CH=$CH_2$ groups with relatively high oxygen content (~7.3wt. %). The mechanical properties improvement led by ameliorated interface was researched. Samples with ameliorated interface employing LPVCS showed better mechanical properties than those employing ploycarbpsilane (PCS). The flexural strength and fracture toughness of the Cf/SiC composites fabricated with PCS were 301 MPa and 11.2 MPa•$m^{1/2}$, respectively, whereas the equivalent values of the samples fabricated with ameliorated interface were 421 MPa and 25.6 MPa•$m^{1/2}$, respectively. Employing LPVCS for the first infiltration and pyrolysis cycle formed suitable interface, then employing PCS as precursor for the left cycles formed relatively low oxygen content matrix, this the main reason of mechanical properties improvement.

## 1. Introduction

Thermal structural materials used in the advanced aero and space applications, such as turbine engine components, hypersonic flight vehicles, and space-craft reentry protection systems, are required to possess extraordinary mechanical properties under high temperature [1-4]. Continuous carbon fiber reinforced SiC matrix composites (Cf/SiC) are considered as one of the most promising thermal structural materials owing to their outstanding high temperature mechanical properties, fracture toughness, and low density [5-7].

There are three common ways to fabricate the Cf/SiC composites: chemical vapor infiltration (CVI), liquid silicon infiltration (LSI), and precursor infiltration and pyrolysis (PIP) [8]. The materials' shape as well as the fabrication time and cost vary for the different fabrication approaches. The CVI process has the longest fabrication period with a porosity of 10-15%. The LSI process requires a shorter fabrication period of 20-30 hours, and the composites have smaller porosity, however the high process temperature will be harmful to the carbon fiber. The PIP process have been widely employed for fabrication of composites because of its advantages in drastic reduction in costs, large-scale production, fabrication with complicated shapes and controllable microstructure.

The coating is important in order to improve the mechanical properties of composites. The mechanical properties of Cf/SiC composites with coating, such as PyC, EBC, carbon nanotube, showed more than twice the flexural strength and fracture toughness [9-11]. The coating was usually prepared by chemical vapor deposition (CVD) method, however CVD process always cost much time.

In the author's preliminary work we found that the precursor LPVCS pyrolyzed products formed suitable interface with T300 carbon fiber, however, the high oxygen content of LPVCS pyrolyzed products reduced the high temperature performance of matrix. Employing LPVCS in the first infiltration and pyrolysis cycle to form suitable interface and then employing PCS in the left cycles to form low oxygen content matrix is a method to improve the mechanical properties of Cf/SiC composites. In this work, the mechanical properties of Cf/SiC composites at room and elevated temperature was studied in detail. In addition, the microstructure were also analyzed.

## 2. Materials and experimental proceduce

### *2.1. Materials*

In the fabrication of composites, T300 carbon fiber was employed as the reinforcement. The general characteristics of the T300 carbon fiber are listed in Table I. All samples were braided into 3D performs by three dimension and four steps braiding technique. The fiber volume fraction and the braiding angle of the composites were 36.3% and 24.8°, respectively. PCS and LPVCS were provided by NUDT, China. The continuous carbon fiber performs were infiltrated with LPVCS in the first cycle and then PCS to the end. Pyrolysis was carried out at 1200 °C in Ar. LPVCS required crosslinking at 300 °C before pyrolysis. The infiltration and pyrolysis process was repeated until the weight increase was less than 1%.

### *2.2. Experimental proceduce*

The densities and open porosities of the Cf/SiC composites were measured by Archimedes' method using kerosene oil as the immersion medium. The bending strength of the composites was characterized by three point bending test following the general guidelines of the Chinese Standard GBT6569-2006. The dimensions of a test samples were 4 mm × 4mm × 65mm, the span length and the cross head speed were 50 mm and 0.5 mm/min, respectively. The fracture toughness of the Cf/SiC composites was determined with the single edged-notched beam method (GBT 23806-2009). The dimensions of the samples for the test were 3.5 mm (B) × 7.0 mm (H) × 40.0 mm (L). The span length was 30.0 mm and the crosshead speed was 0.1 mm/min. Analysis of the pyrolysates was performed by X-ray diffraction (XRD, D8 Advance, Bruker, Germany). The fracture surfaces was observed by scanning electron microscopy (SEM, S-4800, Hitachi, Japan). The oxygen content was measured using an N/O analyzer (EMGA-820, Horiba, Japan) and the carbon content was measured using a C/S analyzer (EMIA-320V2, Horiba, Japan).

**Table 1.** General characteristics of the carbon fiber.

| | Diameter (μm) | Density ($g \cdot cm^{-3}$) | Tensile strength (GPa) | Young Modulus (GPa) |
|---|---|---|---|---|
| T300 | 5.1±0.2 | 1.76 | 6.2±1.9 | 391 |

## 3. Results and discussion

### *3.1. General Characteristics of PCS and LPVCS*

The molecular structures of PCS and LPVCS are shown in Fig. 1. PCS has a main chain of Si-C, with $SiC_4$ and $SiC_3H$ building blocks. LPVCS is the product of LPCS and V4 preliminary reaction. LPCS is Low-molecular polycarbosilane and has a familiar structure with PCS but with low Si-H content. V4 refers to 2, 4, 6, 8-tetravinyl-2, 4, 6, 8-tetramethylcyclotetrasiloxane and has $-CH=CH_2$ and Si-O. When heated to 300 °C, LPCS and V4 form a net structure. LPVCS is a liquid precursor with excellent wettability at room temperature that demonstrates a similar ceramic yield to PCS, with a pyrolysis temperature was 1200 °C. There is more oxygen in the SiCxOy phase when using more LPVCS. The SiCxOy phase reduces the mechanical properties of Cf/SiC composites at elevated temperature.

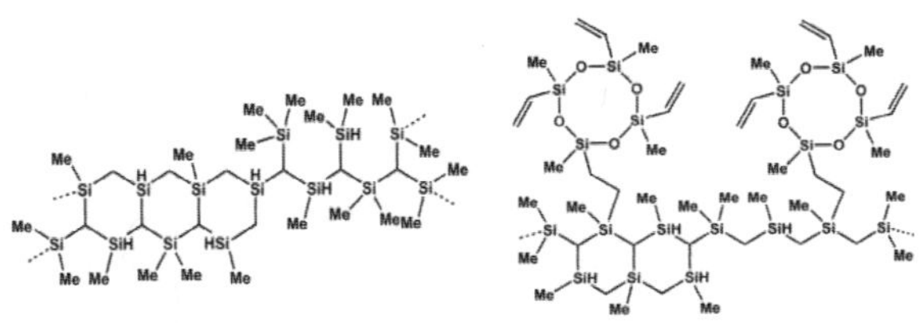


**Figure 1**. Molecular structures of PCS and LPVCS

### *3.2. XRD diffraction patterns*

XRD diffraction patterns XRD diffraction patterns of the pyrolysis products of PCS and LPVCS at 1200 °C are shown in Fig. 2. The pyrolysis products of PCS and LPVCS were almost the amorphous SiC. However, weak peaks appeared at 36.5°, 60.1° and 71.9° associated with the (111), (220) and (311) planes of β-SiC crystals, respectively. A weak peak was observed at 26.5°, which indicated the existence of free carbon.

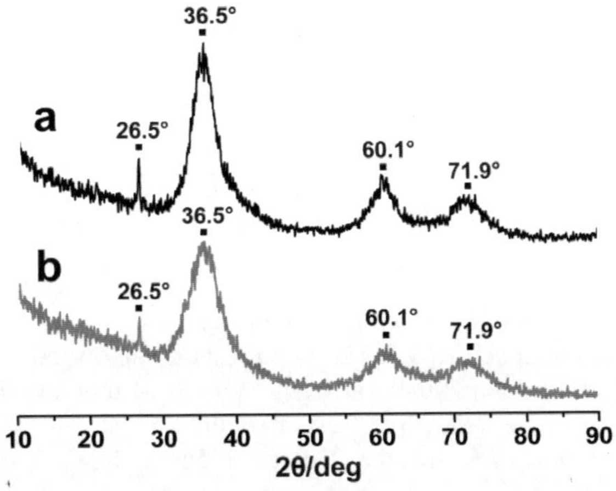


**Figure 2.** XRD diffraction patterns of the pyrolysis products of PCS (a) and LPVCS (b) at 1200 °C

### *3.3. Properties of Cf/SiC Composites at Room and Elevated Temperature*

The weight changes and density increment of the prepared Cf/SiC composites during the PIP process are shown in Fig. 3. The density of Cf/SiC composites increased and weight increase reduced with the infiltration and pyrolysis cycle increase. After 12 cycles of infiltration and pyrolysis, the density and porosity prepared samples were 1.79g/cm$^{-3}$ and 12.1%, respectively.

The mechanical properties of Cf/SiC composites at room and elevated temperature (1400 °C, 1600 °C) are showed in Table II. With ameliorated interface by employing LPVCS for the first cycle of infiltration and pyrolysis, the flexural strength and fracture toughness of the Cf/SiC composites were 421 MPa and 25.6 MPa•m$^{1/2}$, respectively, which were 40% and 128% higher than those (301MPa, 11.2 MPa•m$^{1/2}$) fabricated with PCS. After heat-treatment (1400 °C) for one hour in Ar, the flexural strength and fracture toughness of the Cf/SiC composites were 395 MPa and 22.9 MPa•m$^{1/2}$, respectively, with high retention rate approximately 90%. After heat-treatment (1600 °C) for one hour in Ar, the flexural strength and fracture toughness of the Cf/SiC composites were 297 MPa and 18.2 MPa•m$^{1/2}$, respectively, with retention rate approximately 70%.

In summary, employing LPVCS for the first infiltration and pyrolysis cycle is a method to form suitable interface improving mechanical properties, in addition the samples get high mechanical properties retention rate at elevated temperature because of low oxygen content.

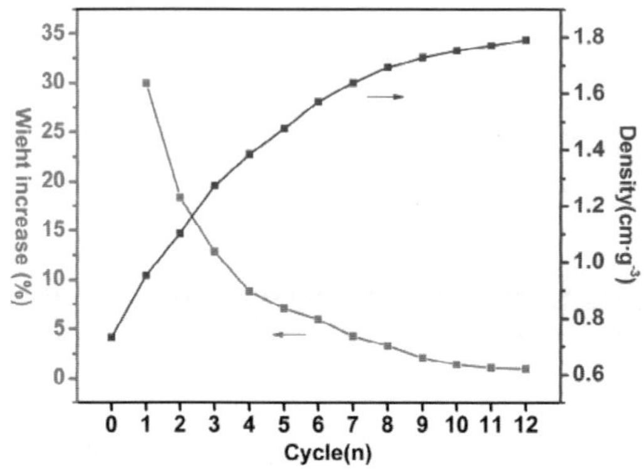


**Figure 3.** Weight changes and density increment of the prepared Cf/SiC composites during the PIP process

**Table 2.** Mechanical Properties of /SiC Composites

| | Flexural strength/MPa | Fracture toughness/MPa·$m^{1/2}$ |
|---|---|---|
| 1200 °C | 421 | 25.6 |
| 1400 °C | 395 | 22.9 |
| 1600 °C | 297 | 18.2 |

*3.4. Microstructural characterization and damage mechanisms*

Typical fractured surfaces of the specimens for bending tests are shown in Fig 4. Toughness fracture formed after ameliorated interface, showing significant amounts of unstick, fracture and pulling out of carbon fiber. T300 carbon fiber and PCS pyrolysis products always formed strong interface force because of groove on the carbon fiber surface. However, LPVCS pyrolysis products formed weak interface with T300 carbon fiber because the oxygen in matrix reacted with the carbon fiber.

In Ar, three reasons led to the damage in mechanical properties. First, oxidation of the carbon fiber led to strength reduce (Eqs. (1)), damage to the surface of the carbon fiber was observed in Fig 5. Second, decomposition of the SiCxOy phase led to porosity and non-densification of the matrix (Eqs. (2)), which was observed in Fig 5. Third, growth of the β-SiC grain size reduced the matrix strength, which was observed in Fig 6.

$$C\ (s) + O_2\ (g) = CO_2\ (g) \quad (1)$$

$$SiCxOy \rightarrow SiC\ (s) + C\ (s) + SiO\ (g) + CO\ (g) \quad (2)$$

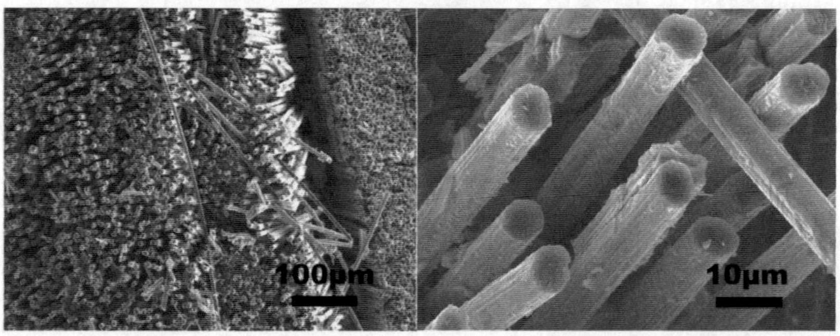


**Figure 4.** Typical fractured surfaces of the specimens

**Figure 5.** Typical fractured surfaces of the specimens after heat-treatment

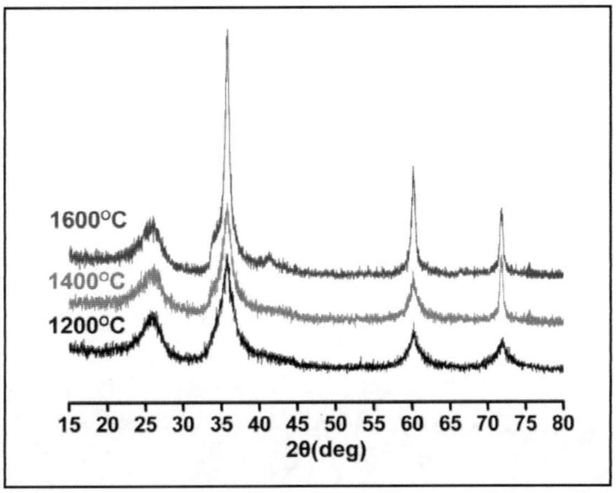


**Figure 6.** XRD diffraction patterns of samples after heat-treatment

## 4. Conclusions

Employing LPVCS for the first infiltration and pyrolysis cycle is a method to form suitable interface to improve the mechanical properties of Cf/SiC composites, in addition the samples get high mechanical properties retention rate at elevated temperature because of low oxygen content.

In Ar, three reasons led to the damage in mechanical properties. First, oxidation of the carbon fiber led to strength reduce. Second, decomposition of the SiCxOy phase led to porosity and non-densification of the matrix. Third, growth of the β-SiC grain size reduced the matrix strength.

The further work will focus on reducing the porosity of Cf/SiC composites.

## 5. Acknowledgments

The study was supported by the National Nature Science Foundation of China (No. 51403233), the aid program for Science and Technology Innovative Research Team in Higher Educational Institutions of Hunan Province and the aid program for Innovative Group of National University of Technology. The authors are also thankful for The Fund of Science and Technology on Advanced Ceramic Fibers and Composites Laboratory (No. 9140A12040215XX01001).

## 6. References

[1] Y. Katoh, L.L. Snead, C.H. Henager, A.Hasegawa, A.Kohyama, B.Riccardi, 2007 *J. Nucl. Mater.* 659-671.

[2] T. Pichon, R. Barreteau, P. Soyris, 2009 *Acta Astronaut* 165-176.

[3] T. Nozawa, T. Hinoki, A. Hasegawa, A.Kohyama, Y. Katoh, L.L. Snead, 2009 *J. Nucl. Mater.* 622-627.
[4] S. Fouquet, M. Rollin, R. Pailler, 2008 *Wear* 850-856.
[5] A.G. Odeshi, H. Mucha, B. Wielage, 2006 *Carbon* 1994-2001.s.
[6] Y. Yorozu, M. Hirano, K. Oka, and Y. Tagawa, 1987 *J. Magn. Japan* p 301.
[7] Y. Xiang, W. Li, S. Wang, 2012 *Surf. Coat. Technol* 197-202.
[8] Y Katoh, M Kotani, H Kishimoto, 2001 *J. Nucl. Mater* 42-47.
[9] Z Luo, X Zhou, J Yu, 2014 *Surf. Coat. Technol* 146-153.
[10] B Yang, X.G. Zhou, J.S. Yu, 2015 *Mater, Sci, Forum* 186-191.
[11] S Zhao, Z Yang, X Zhou, 2015 *J. Am. Ceram. Soc* 1332-1337

IOP Conf. Series: Materials Science and Engineering **378** (2018) 012009 doi:10.1088/1757-899X/378/1/012009

# Study on the Compressive Mechanical Properties of HTPB Propellant at Low Temperature

**Xiangdong Chen[1,2,*], Xin-long Chang[1], Youhong Zhang[1], Lei Zhang[1]**

[*]Corresponding author: chen987dong@163.com

[1]Xi'an High-tech Institute, Xi'an 710025, shannxi, China

[2]96630 Unit of PLA, Beijing 102206, China

**Abstract**. To study the low temperature effects of compressive mechanical properties on Hydroxyl-terminated polybutadiene (HTPB) propellant, a quasi-static mechanical experiment was conducted. The results show that compressive mechanical parameters are closely related to strain rate and low temperature. With the decrease of temperature and increase of strain rate, the modulus and compressive strength of HTPB propellant increase obviously. Based on the time-temperature equivalence principle (TTEP), the master curves of compressive strength and initial modulus for HTPB propellant were obtained, which can facilitate the structural integrity analysis of the propellant.

## 1. Introduction

Hydroxyl-terminated polybutadiene(HTPB) propellant is a kind of energetic material filled with a certain amount of ammonium perchlorate particles and aluminum particles. HTPB propellant is widely used in solid rocket motors (SRMs) for the propulsion of missile and space transportation because of its high energetic characteristics and good mechanical properties. The compressive mechanical properties of HTPB propellant are extraordinary important for structural integrity of solid rocket motor grain. Furthermore, SRMs are often required to operate in cold areas, which can cause significant changes in mechanical properties of the propellant. So it is necessary to research the compressive mechanical properties at low temperature [1-2].

It is important to fully understanding of mechanical properties on energetic material at different conditions. Himanshu [3-4] examined the temperature effects and aging time on mechanical properties of solid rocket propellants. He found that any change in temperature and aging time brought significant change in the tensile strength, percentage elongation, and elastic modulus of the propellant. Thompson et al. [5] conducted compressive and tensile testing for plastic-bonded explosives(PBXs) 9501 and 9502, spanning a wide range of strain rates and temperatures. Williamson et al. [6] measured the compressive strength of the energetic composition EDC37 at a strain rate of $10^{-3}$/s over a range of temperatures from 208 K to 333 K. Based on the experimental results, the failure stress was found to exhibit a similar trend with increase of the logarithm of strain rate and decrease of temperature. Riadh et al. [7] investigated viscoelastic response of high-density polyethylene(HDPE) under long-term tensile and compressive creep. Yang et al. [8] examined compressive mechanical testing of HTPB propellant at room temperature over strain rate ranging from $1.7\times10^{-4}$/s to 2500/s. And he proposed a rate-dependent constitutive model to described the mechanical behaviour. Wang et al. [9] studied the biaxial tensile stress responses of HTPB propellant at room temperature, and they [10] also conducted the uniaxial tensile tests at low temperatures.

To sum up, most scholars mainly focused on the energetic material tensile properties or the properties at room temperature, but there are relatively little related reports in compressive mechanical properties of HTPB propellant at low temperature. In this paper, the compressive properties were investigated at low temperature, and the master curves of compressive mechanical properties for HTPB propellant were obtained. Our goal is to study the compressive mechanical behaviours of HTPB propellant, which will facilitate the structural integrity analysis.

## 2. Experimental part

### *2.1 Materials and sample configuration*

The HTPB propellant investigated here was manufactured by casting method consisting of 86% by weigh AP power and Al power, 10% by weigh HTPB binder and 4% by weigh other liquid additives. The propellant was machined into small cylindrical specimens of both diameter and height 20mm (Figure.1). All specimens were machined from the same manufacturing batch, and then stored in a desiccator cabinet at 40°C for 24h before testing to remove any residual stresses and eliminate any variability caused by humidity.

**Figure 1.** Geometrical shape and the specimen

### *2.2 Experiment condition design*

The experimental temperatures were 25°C, -10°C, -20°C,-30°C, -40°C and -50°C. The experimental compression speed of specimens was about 4 mm/min（$1/300s^{-1}$）, 10mm/min（$1/120s^{-1}$）, 40mm/min（$1/30s^{-1}$） and 100mm/min（$1/12s^{-1}$）. Each test was repeated at least five times to measure repeatability of the experimental results. The temperature change was realized by the temperature controller. The specimens were stored for at least 1h at the specified temperature before test.

## 3 Results and discussion

### *3.1 Stress-strain curve*

(1) The experimental stress-strain data is transformed into true stress and strain. The typical curves of at 25℃ and -30℃ are represented in Figure 2. Because of the viscoelastic material nature, temperature and strain rate greatly affect the mechanical behaviour of HTPB propellant. With the decrease of temperature and the increase of strain rate, the stress increase obviously. In addition, the curves show strong nonlinear characteristics under various condition.

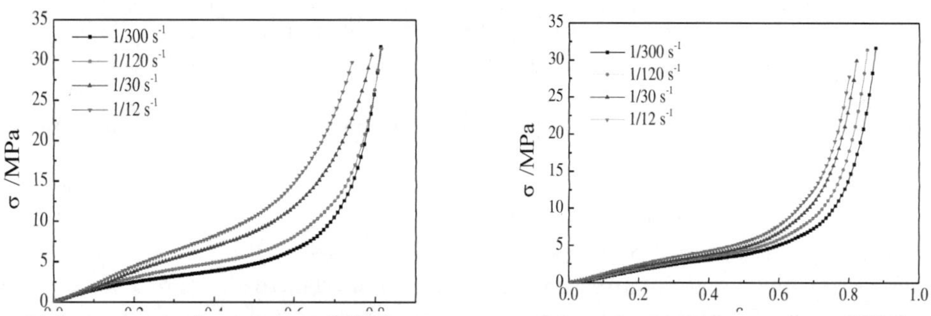


(a) stress-strain curve at 25°C (b) stress-strain curve at -30°C

**Figure 2.** typical stress-strain curves

(2) By analyzing the curves of Figure 2, the curve can be divided into three regions, and the characteristics of the curve are showed in Figure 3. The three regions are initial linear elastic region, strain hardening region and stress failure region. In elastic region (region I), the stress is linearly related to strain, and the slope is defined as the initial elastic modulus, and the damage of the specimen is not particularly severe at this stage. In strain hardening region (region П), the stress increases slowly as the strain increases. The middle part of the specimen is subjected to circumferential tensile stress. With the increase of strain, the cracks around the specimens obviously expand, and the greater the strain rate and the lower the temperature, the greater crack growth rate. In stress failure region (region III), the slope of the curve increases rapidly, which means the specimen has failed. In Figure 3, the A and B point represents the boundaries of each phase, where the B point means the compressive yield strength. The analysis of the mechanical properties is based on the definition of modulus and compressive yield strength.

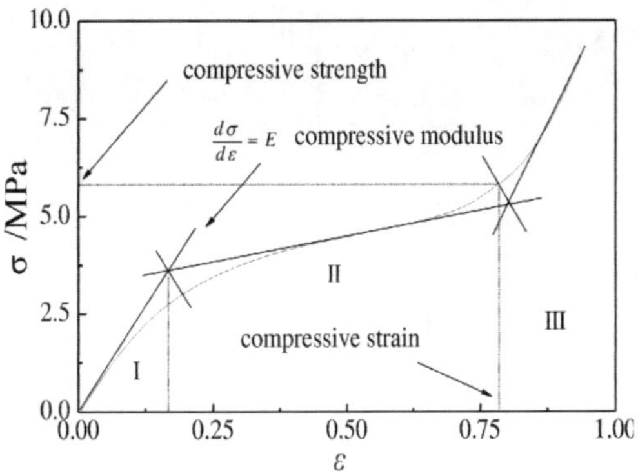


**Figure 3.** characteristics of stress-strain curve

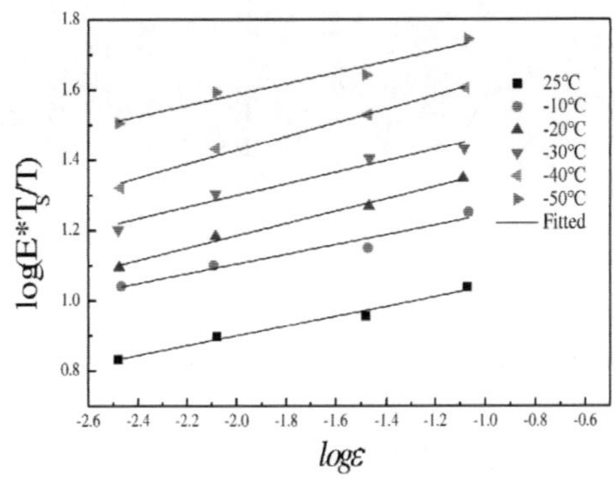


**Figure 4.** modulus vs. logarithmic strain rate at various test temperatures

*3.2 Mechanical properties and master curve*

Figure 4 indicates that the modulus is linearly related to the logarithm of strain rate, with a higher slope at lower temperatures. The modulus at -50°C and $1/12s^{-1}$ is 33.5MPa, which is 3.41 times of the value at 25°C and $1/300s^{-1}$. The interaction of strain rate and low temperature strongly influence the change of modulus.

By shifting the modulus value horizontally along the logarithmic strain rate axis with the reference temperature of 25°C, we can get the modulus shift factor $\alpha_T$ curve represented in Figure 5. By means of Figure 4 and Figure 5, the compressive initial modulus master curve can be obtained in Figure 6.

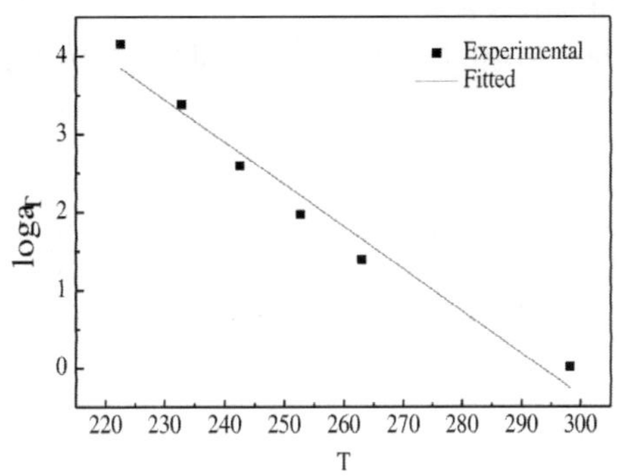


**Figure 5.** modulus logarithmic shift factor vs. temperatures

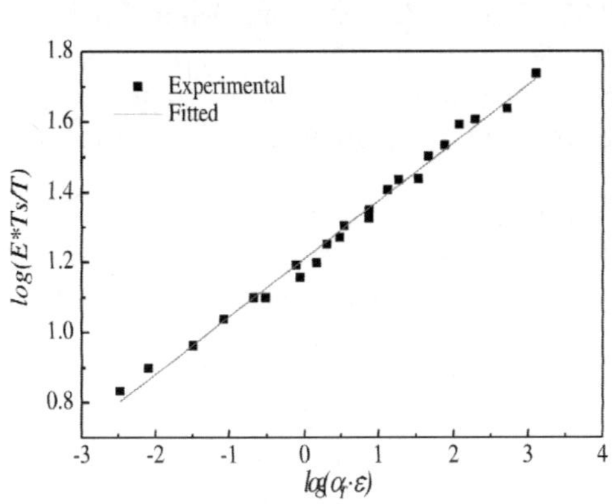


**Figure 6.** the master curve for compressive modulus

Using the same way above, we also obtain the compressive strength master curve (Figure 7) with the reference temperature of 25°C. And the strength shift factor $\alpha_T$ is inserted in Figure 7. Table1 summarizes two equations, which describe the relationship of mechanical parameters with the reduced strain rate $\alpha_T \cdot \dot{\varepsilon}$ .

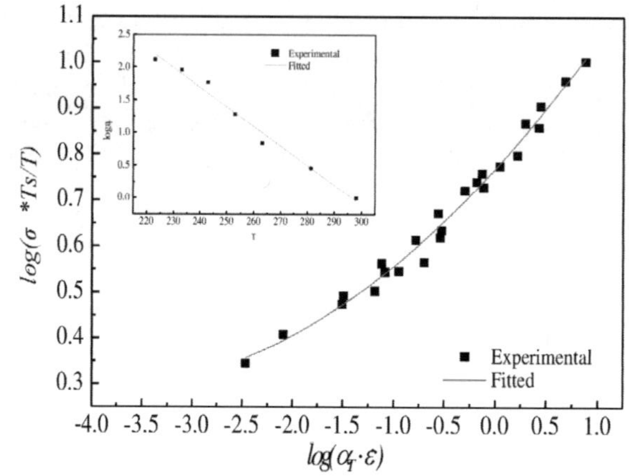


**Figure 7.** the master curve for compressive strength

**Table 1** Equations for mechanical parameters

| Mechanical parameters | Fitted formulas | Correlation coefficient/$R^2$ |
|---|---|---|
| Compressive strength | $log(\sigma \cdot T_S / T) = 0.77 + 0.24 \cdot log(a_T \cdot \dot{\varepsilon}) + 0.03 \cdot log^2(a_T \cdot \dot{\varepsilon})$ | 0.98 |
| initial modulus | $log(E \cdot T_S / T) = 1.2 + 0.16 \cdot log(a_T \cdot \dot{\varepsilon})$ | 0.99 |

## 4. Conclusions

In this paper, the mechanical behaviours of HTPB propellant under low temperatures and quasi-static compression were investigated. From the experiment, we can draw the following conclusions.

(1) Results show that HTPB propellant are closely related to temperature and strain rate. The stress and modulus change more obviously at lower temperature and higher strain rate.

(2) The master curves of mechanical parameters and their expressions are shown in this investigation. The master curves of compressive strength and initial modulus are expressed as the quadratic and linear function of the reduced strain rate, which can provide the theoretical basis for further analyzing the structural integrity of propellant grain.

## 5. References

[1] Structural assessment of solid propellant grains[R]. AGARD-AR-350, France, Advisory Group for Aerospace Research & Development, 1997.

[2] Amos R J. On a viscoplastic characterisation of solid propellant and the prediction of grain failure on pressurisation cold[C]. 37th AIAA/ASME/SAE/ASEE Joint Propulsion Conference and Exhibit, July 2001, Salt Lake City, Utah. AIAA 2001-3719: 1-10.

[3] Himanshu Shekhar. Effect of Temperature on Mechanical Properties of Solid Rocket Propellants[J]. Defemce Science Journal, 2011, 61(6): 529-533.

[4] Himanshu Shekhar Studies on Stress-Strain Curves of Aged Composite Solid Rocket Propellants[J]. Defence Science Journal, 2012,62(2) :90-94.

[5] Darla Graff Thompson, Racci Deluca, Geoff W. Brown. Time-Temperature Analysis, Tension and Compression in PBXs[J]. Journal of Energetic Materials, 2012, 30: 299-323.

[6] Williamson D M, Siviour C R, Proud W G, et al. Temperature-time response of a polymer bonded explosive in compression[J]. Journal of Physics D: Applied Physics, 2008, 41(8): 1-10.

[7] Rladh Elleuch, Wafa Taktak. Viscoelastic Behavior of HDPE Polymer using Tensile and Compressive Loading[J]. 2006, 15,111-116.

[8] Yang Long, Xie Kan, Pei Jiang-feng, et al. Compressive mechanical properties of HTPB propellant at low,intermediate, and high strain rates[J]. J. APPL. POLYM. SCI. 2016, 133(23): 1-9.

[9] Wang Zhe-jun, Qiang Hong-fu, Wang Guang, et al. A new test method to obtain biaxial tensile behaviors of solid propellant at high strain rates[J]. Iran Polym J, 2016, 25(6): 515-524.

[10] Zhejun Wang, Hongfu Qiang, Guang Wang. Experimental Investigation on High Strain Rate Tensile Behaviors of HTPB Propellant at Low Temperatures[J]. Propellants Explos. Pyrotech. 2015, 40, 814-820.

IOP Conf. Series: Materials Science and Engineering **378** (2018) 012010 doi:10.1088/1757-899X/378/1/012010

# A Study of Impact Energy on Eggshell Crack

**Po-Ting Chou**[1], **Jeng-Liang Lin**[2*]
[1]Dept. Biomechatronic Engineering, National Chiayi University, Taiwan
[2]Dept. Biomechatronic Engineering, National Chiayi University, Taiwan

[*]Corresponding Author, Tel +886-5-2717656, Email: jllin@mail.ncyu.edu.tw

**Abstract.** The lengthy and irregular cracks occur when eggs are impacted. These egg samples with cracks can be used for thermal image to study the display of the crack. The purpose in the study was to analyse eggs impacted by free fall and measure the crack length with a self-developed image analysis system, further investigating the relationship between the impact energy and the crack length. The image analysis system comprised Image Import, Gray Level Processing, Binary Transform, Dilation and Labelling. Finally, all crack blocks could be obtained, as well as the area, perimeter and length could be determined. The test results showed that a height of free fall between 24-42 mm or an impact energy of 12-26 mJ would cause cracks similar to those when eggs were collected. The larger energy was, and the longer crack length was, presenting in linear relationship, as well as the egg source to the relation would be greatly influenced.

## 1. Introduction

The egg collection and processing system comprised egg feeding, egg washer, dryer, and post inspection. If some cleaned and dried eggs were not cleaned or damaged, they must be picked out by the post inspection to ensure the quality of egg. The final purpose of this study was to establish a set of wash and dry procedures associated with the egg-wash operation, where the temperature difference in the cooling and heating processes was acquired by thermal image to analyse the crack pattern of eggshell and display effect in thermal image, offering another choice from the sound detection [1-2]. A crack manually produced on the surface of eggshell was the first step in the study. Because the impact is one of major factors to produce the cracks on eggshell, the impact of free fall on eggs can simulate cracks when they are collected. After they suffered from the impact energy, the lengthy and irregular cracks would be formed. These samples could be used in the tests of crack detection by thermal image [3]. The purpose of this study was to analyse the impact of free fall on eggs, influence the crack pattern from different heights, as well as measure the crack length by the self-developed image analysis system to further investigate the relationship between the impact energy and the crack length.

IOP Conf. Series: Materials Science and Engineering **378** (2018) 012010 doi:10.1088/1757-899X/378/1/012010

## 2. Materials and Methods

The crack impact test was conducted in the manners of eggs falling freely to hit the hard board, where the height of free fall (about 2-4cm) was accurately adjusted by a plain paper and the bottom was a hard plate in horizon. The impact energy that eggs rolled from the top of paper to the lower plate would crack eggshell. Eggs in the test were purchased from a supermarket and cracked by the impact energy of free fall from four heights. 12 eggs were required and labelled as A01-A12 when repeating three times from every height. In order to test the influence of eggs from the different sources, 8 eggs were purchased from the other supermarket to crack in free fall from the same heights. All heights were repeated for two times, and eggs were labelled as B01-B08.

The way to test the impact energy was free fall from the height. Stacking paper in different quantities can simply produce the height difference. There were four heights of free fall in the study, which are 24, 28, 34 and 42 mm. The height of free fall selected between 24-42 mm was because the results of preliminary test showed that it was difficult to observe the cracks from a height less than 24 mm, and the height greater than 42 mm would produce large cracks and outflow egg liquid.

The crack length analysed by image process needs to shoot the eggshell cracks in a specific way, so that can obtain the clear images to accurately measure the crack length. The structure to shoot the crack images was a plain plastic box with a depth of 12 cm, where a piece of black paper was laid on the bottom and two supports were fixed on the top to load the camera. The side of a cracked egg upwardly faced, so that the eggshell image in black background could be obtained at the fixed height, as shown in figure 1.

**Figure 1.** Image acquiring setup for cracked egg

The major axis and the minor axis of an egg were measured by a caliper before test, so that the length of every pixels in the image could be determined during the image process, and then weighed by an electronic scale to calculate the impact energy by potential energy formula (W = mgh). Afterward, eggs were loaded on the edge of stacked paper and lightly free fallen to ensure other forces were considerably small except for gravity. One or two images then were shot when every egg fell from a height to crack. In order to match a resolution of 320 × 240 for thermal imager, the image resolution for the image analysis system was set as 320 × 240. Because tiny cracks on eggshell might not appear in the image due to the original images being directly imported to the image analysis system, a marker was used to draw black lines on the cracks. Therefore, the cracks could be clearly appeared on the image analysis system.

## 3. Image Analysis System

Plant characteristics, such as leaf shape and area, have been studied numerously with image analysis [4-7]. Crack length on eggshell was determined by image analysis that was programmed by Visual Basic, where the interface form is shown in figure 2.

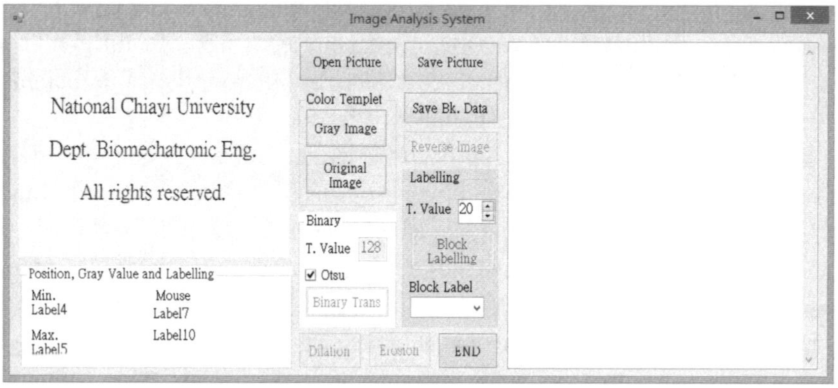


**Figure 2.** Interface of image analysis system

The image analysis system comprised Image import, gray level processing, binary transform [8], dilation and labelling [9]. An image of cracked eggshell drawn by a marker was selected from the Open Picture button in the interface. The width-height ratio of an image was adjusted as 4:3 before image selection. Otherwise, the image would be compressed and deformed. The image at the right side of the figure 3 shows the distorted image after image import.

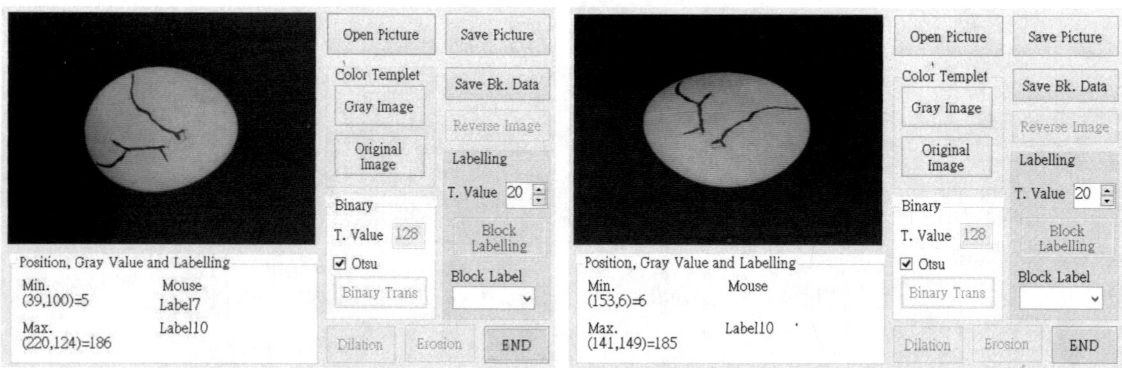


**Figure 3.** True (left) and distorted (right) images after image import

Using the Gray Image button transformed the color image to the gray-level image, and then using the Binary Trans button transformed the image to the black and white image, where eggshell without and with cracks were presented in white and in black respectively (as shown in figure 4).

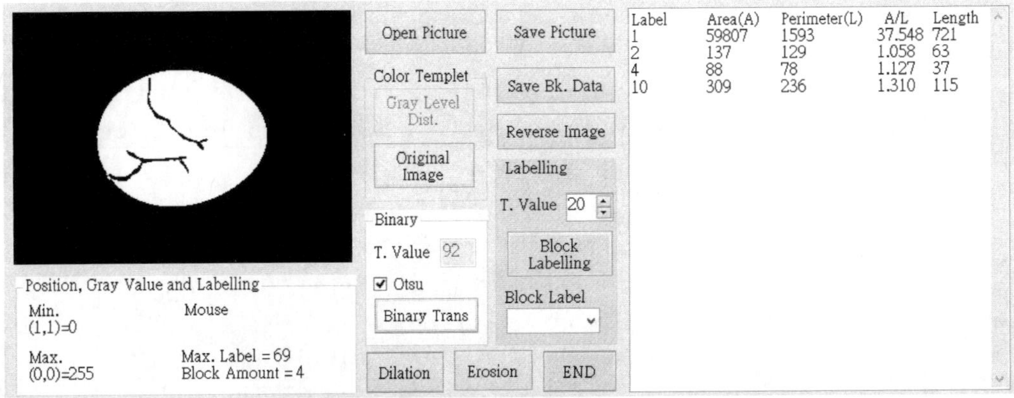


**Figure 4.** Analysis screen of crack length

After pressing the Binary Trans button, the Block Labelling button was pressed to label all blocks in black by labelling as well as analyze the area, the perimeter and the length, where the area was the total number of pixel in the block and the perimeter was the total number of pixel on the block edge. However, if the edge pixels were the outer corner ones, the values were multiplied by 1.414 so that a perimeter of the block could be accurately obtained. When performing the image analysis, the Dilation button could be pressed to dilate the crack block. Because the cracks were usually too thin, they had to be dilated in order to prevent determining as different blocks by labelling when blocks connected diagonally. The dilated crack was basically a line in a width of 3 or 5 pixels. If the factors at the end portion were eliminated, the length was about a half of the perimeter.

Because the length unit in the image analysis system was pixel and the length unit of the true crack was mm, the unit must be converted. First of all, the measurements of egg length (mm) divided by the length of eggshell image (pixel) represented the length of every pixel (mm/pixel), and then the unit conversion multiplied by the crack length on eggshell (pixel) could then obtained the actual crack length on eggshell (mm).

## 4. Results and Discussions

The basic physical properties and results from image analysis of eggs under test are shown in table 1. The major (meridional) axis and the minor (circumferential) axis in the first batch of eggs (A) were between 55-59 mm and 43-45 mm respectively, where the average values were 56.4 mm and 43.6 mm. The weight was in the range of 54-63 g, where the average value was 60.3 g. The major (meridional) axis and the minor (circumferential) axis in the second batch of eggs (B) were 54-61 mm and 43-45 mm respectively, where the average values were 57.6 mm and 43.9 mm. The weight was in the range of 56-66 g, where the average value was 60.7 g. Although the eggs in two batches came from the different sources, basically they had the same size and weight. When the height of free fall was 24-42 mm, the impact energy was 12-26 mJ. As for the length represented by each pixel, the test results showed that there were somewhat different between horizontal and vertical directions. The reason was that the orientation of major axis was not located on the true direction of horizon when shooting the egg image. The maximum and minimum differences in the first batch of eggs (A) were 8.4% for A05 and 0.3% for A02, where the average difference was 2.5%. The maximum and minimum differences in the second batch of eggs (B) were 2.5% for B04 and 0.3% for B06, where the average difference

IOP Conf. Series: Materials Science and Engineering **378** (2018) 012010 doi:10.1088/1757-899X/378/1/012010

was 1.2%. The average value of length represented per pixel for both horizontal and vertical direction was 0.325 mm/pixel.

**Table1.** Experimental averages of each height of free fall for the two batch eggs

| Egg # | Free Fall Height (mm) | Crack Length (mm) | Major axis (mm) | Minor axis (mm) | Weight (g) | X resolution (mm/pixel) | Y resolution (mm/pixel) | Impact energy (mJ) |
|---|---|---|---|---|---|---|---|---|
| A01~A03 | 24 | 31.54 | 56 | 44 | 56.77 | 0.324 | 0.320 | 13.36 |
| A04~A06 | 28 | 47.43 | 56 | 44 | 61.14 | 0.339 | 0.326 | 16.79 |
| A07~A09 | 34 | 55.27 | 57 | 44 | 62.65 | 0.325 | 0.322 | 20.89 |
| A10~A12 | 42 | 65.61 | 56 | 43 | 60.48 | 0.334 | 0.326 | 24.92 |
| B01~B02 | 24 | 96.47 | 59 | 44 | 61.64 | 0.323 | 0.323 | 14.25 |
| B03~B04 | 28 | 81.36 | 59 | 44 | 63.18 | 0.324 | 0.317 | 17.35 |
| B05~B06 | 34 | 96.47 | 56 | 44 | 59.59 | 0.321 | 0.324 | 19.91 |
| B07~B08 | 42 | 163.37 | 57 | 44 | 58.37 | 0.326 | 0.327 | 24.05 |

Some of the physical crack images after eggs were impacted by free fall are shown in figure 5. Eggs free falling from a height of 2.4 cm (A01) would cause shorter cracks. Eggs free falling from a height of 2.8 cm (A04) would cause longer cracks, but would not outflow egg white. Eggs free falling from a height of 3.4 cm (A07) would cause longer cracks. Although cracks were deep a little bit, egg white would not outflow. Eggs free falling from a height of 4.2 cm (A11) would cause even longer cracks, whereas some egg white would outflow. Therefore, for an impact test in free fall, a height should not be greater than 4.2 cm. In the impact of free fall from the same height, the cracks in the second batch of eggs appeared more obvious than those in the first batch of eggs to indicate that the eggshell strength was greatly related to the sources, which could be related with the chicken breed, feed and environmental management. When eggs free fell, the circumferential area would directly suffer from the impact energy. Since the strength on the circumferential area was smaller than that on two meridional areas [10], the cracks were likely located in the circumferential area. In some tests with higher impact energy, the cracks on eggshell could extend to the reverse side of the impact. If this result occurred, two crack images would be required to shoot so that the crack length on eggshell could be determined completely.

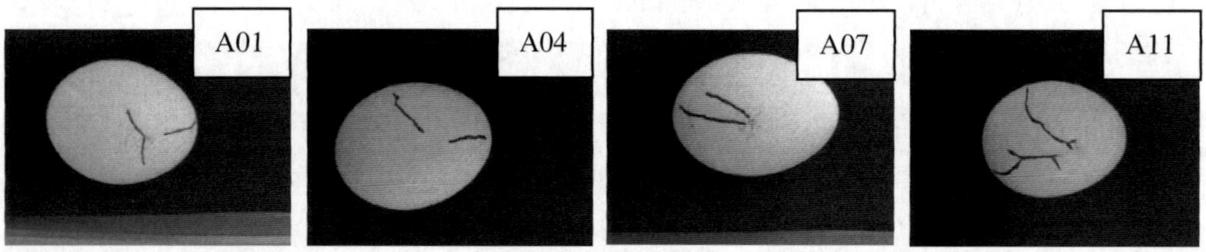


**Figure 5.** Physical crack images after eggs being impacted by free fall

Height of free fall had great influence on cracks of eggshell (figure 5), where the higher height was and the cracks would be more obvious. However, impact energy was the factor to cause cracks on eggshell. In consideration of height in free fall and egg weight, variation of the impact energy on the crack length (figure 6) could be obtained. The results showed that eggs under an impact energy of 12-26 mJ would crack. The larger energy was and the crack would be longer in a linear relationship.

IOP Conf. Series: Materials Science and Engineering **378** (2018) 012010 doi:10.1088/1757-899X/378/1/012010

Although the crack length was in a linear relationship with the impact energy, the relation would show the large difference owing to the different sources. The slope in the first batch of eggs (A) with the impact energy in a linear relationship of crack length was 2.95 mm/mJ, and the slope in the second batch of eggs (B) was 9.02 mm/mJ.

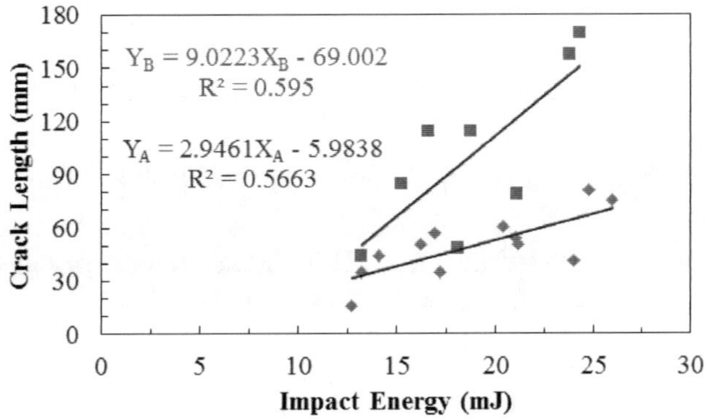


**Figure 6.** Influence of impact energy on crack length, where the data in blue diamond and red square were eggs in the first (A) and second (B) batches

## 5. Conclusions

The self-developed image analysis system in this study could measure the crack length, and further investigate the relationship between the impact energy and the crack length by free fall impact. Based on the test results, specific conclusions are as follows.

The cracks produced from a height of free fall between 24-42 mm could simulate the crack on the eggshells when eggs were collected.

The self-developed image analysis system could analyse the length per pixel in the image by the measurement of the egg size and the appropriate image shooting, where the average value of the length per pixel in the image was 0.325 mm. Then, the crack length on eggshell could be determined by the pixels of the crack edge in the image.

Eggs would produce the cracks when suffering from an impact energy of 12-26 mJ. The larger energy was and the crack would be longer in a linear relationship and related with the egg source.

## 6. References

[1] Li C, Chang J, Cheng C. and Hsieh L. 2011. A Novel Non-Destructive Technology for Inspecting Eggshell Cracks Using Impulsive Response Time. Food Sci. Technol. Res. 17 (1): 1-10

[2] Chen Y-C, Hu M-L and Cheng C-W. 2011. Applying Non-Destructive Techniques to Inspect Preserved Egg Products by Decay Rates. Journal of Food Engineering. 104: 30–35.

[3] Shen P-N, Lei P-K, Liu Y-C, Haung Y-J and Lin J-L. 2016. Development of a temperature measurement system for a broiler flock with thermal imaging. Engineering in Agriculture, Environment and Food. 9 (3): 291-295.

[4] Meyer G E and Davison D A. 1987. An electronic image plant growth measurement system. Transactions of the ASAE. 30(1): 242-248.

[5] Ghazanfari A, Irudayaraj J, Kusalik A and Romaniuk M. 1997. Machine vision grading of pistachio nuts using Fourier descriptors. J. Agric. Engng. Res. 68: 247-252.

[6] Kuo T-Y, Chung C-L, Chen S-Y, Lin H-A and Kuo Y-F. 2016. Identifying rice grains using image analysis and sparse-representation-based classification. Computers and Electronics in Agriculture. 127: 716-725.

[7] Van Henton E J and Bontsema J. 1995. Nondestructive crop measurement by image processing for crop growth control. J. Agric. Engng. Res. 61: 97-105.

[8] Otsu N. 1979. A threshold selection method from gray-level histograms. IEEE Transaction on System, Man, and Cybernetics. 9(1): 62-66.

[9] Haralick R M and Shapiro L G. 1992. Computer and Robot Vision (Volume I). New York: Addison-Wesley.

[10] Lin J, Puri V M and Anantheswaran R C. 1995. Measurement of eggshell thermal-mechanical properties. Transactions of the ASAE. 38: 1769-1776.

IOP Conf. Series: Materials Science and Engineering **378** (2018) 012011 doi:10.1088/1757-899X/378/1/012011

# Vibration and Mode Shape Analysis of Functionally Graded Nanocomposite Plates Reinforced by Aggregated Carbon Nanotube

**Vahid Tahouneh**
Young Researchers and Elite Club, Islamshahr Branch, Islamic Azad University, Islamshahr, Iran.

E-mail: vahid.th1982@gmail.com/vahid.tahouneh@ut.ac.ir

**Abstract.** In the present work, by considering the agglomeration effect of single-walled carbon nanotubes, free vibration characteristics of functionally graded (FG) nanocomposite plates is presented. The volume fractions of randomly oriented agglomerated single-walled carbon nanotubes (SWCNTs) are assumed to be graded in the thickness direction. To determine the effect of CNT agglomeration on the elastic properties of CNT-reinforced composites, a two-parameter micromechanical model of agglomeration is employed. The effects of geometrical and material parameters on the frequency parameters of the FG nanocomposite plates are investigated. For an overall comprehension on 3-D vibration of rectangular plates, some mode shape contour plots are reported in this research work.

## 1. Introduction

Nowadays, the use of carbon nanotubes in polymer/carbon nanotube composites has attracted wide attention [1]. A high aspect ratio, low weight of CNTs and their extraordinary mechanical properties (strength and flexibility) provide the ultimate reinforcement for the next generation of extremely lightweight but highly elastic and very strong advanced composite materials. On the other hand, by using of the polymer/CNT composites in advanced multilayered composite materials (sandwich structures) we can achieve structures with low weight, high strength and high stiffness in many structures of civil, mechanical and space engineering. Yas and Tahouneh [2] investigated the free vibration analysis of thick FG annular plates on elastic foundations via differential quadrature method based on the three-dimensional elasticity theory. Tahouneh et al. [3] studied free vibration characteristics of continuous grading fiber reinforced (CGFR) annular plates resting on elastic foundations using DQM. Tahouneh [4] studied the free vibration analysis of thick nanocomposite plates. The author used equivalent continuum model to estimate the material properties of the structure. Tahouneh and Naei [5] used 2D differential quadrature to investigate free vibration and vibrational displacements of elastically supported laminated curved panels with power-law distribution FG layers and finite length. The same authors [6] have considered a 2-D six-parameter power-law distribution for three-dimensional dynamic analysis of thick multi-directional functionally graded rectangular plates with two edges simply supported while the other two edges assumed to be free, clamped and simply supported. Cheng and Batra [7] have studied the buckling and steady state vibrations of a simply supported functionally graded polygonal plate based on Reddy's plate theory. Zenkour [8] presented a two dimensional solution to study the bending, buckling and free vibration of simply supported FG ceramic–metal sandwich plates. Cheng and Batra [9] used first and third order shear deformation theories to relate deflections of a simply supported functionally graded polygonal plate. Reddy [10] presented static and dynamic analysis of the FGM plates based on third order shear deformation theory and by using the theoretical formulation and finite element models. Matsunaga [11] studied vibration analysis of FG plates using 2D higher-order deformation theory. In this work, three-dimensional elasticity theory is used to investigate the free vibration analysis of simply supported functionally graded nanocomposite plates reinforced by aggregated SWCNTs. The applied nanocomposite is assumed a mixture of CNTs (that are randomly oriented and locally aggregated into some clusters) that are embedded in a polymer. The material properties of the nanocomposites plates are assumed to vary along the thickness of plate and estimated through the Mori-Tanaka method. The effects of CNT volume fraction, CNT distribution, cluster distribution and geometric dimensions of plate are investigated on the natural frequency of the FG-CNTRC plates.

## 2. Material Properties in FG-CNTRC Reinforced Composite

### *2.1. Completely Randomly Oriented CNTs*

Consider a CNTRC is made from a mixture of SWCNT (that randomly oriented and locally aggregated into some clusters) and matrix which is assumed to be isotropic. Many studies have been published each with a different focus on mechanical properties of polymer nanotube composites. However, the common theme seems to have been enhancement of Young's modulus. In this section, the effective mechanical properties of the CNT reinforced composite that straight CNTs are oriented randomly or local aggregated in to some clusters are obtained based on the Eshelby-Mori-Tanaka approach. The resulting effective properties for these CNT reinforced composite are isotropic, despite the CNTs are transversely isotropic.When CNTs are completely randomly oriented in the matrix, the composite is then isotropic, and its bulk modulus K and shear modulus G are derived as [12]:

$$K = K_m + \frac{f_r(\delta_r - 3K_m\alpha_r)}{3(f_m + f_r\alpha_r)}, G = G_m + \frac{f_r(\eta_r - 2G_m\beta_r)}{2(f_m + f_r\beta_r)} \tag{1}$$

where subscripts "m" and "r" are referred to matrix and CNT respectively, "f" is volume fraction, $K_m$ and $G_m$ are the bulk and shear moduli of the matrix, respectively. Parameters, $\delta_r$, $\alpha_r$, $\eta_r$ and $\beta_r$ have kind of relation with Hill's elastic moduli for the reinforcing phase (CNTs), [12]. The effective Young's modulus E and Poisson's ratio of the composite is given by:

$$E = \frac{9KG}{3K + G}, \upsilon = \frac{3K - 2G}{2(3K + G)} \tag{2}$$

### *2.2. Effect of CNT Aggregation on the Properties of the Composite*

The CNTs were arranged within the matrix in such manner so as to introduce clustering. It has been observed that, due to large aspect ratio (usually >1000), low bending rigidity of CNTs and Van der Waals forces, CNTs have a tendency to bundle or cluster together making it quite difficult to produce fully-dispersed CNT reinforced composites. The effect of nanotube aggregation on the elastic properties of randomly oriented CNTRC is presented in this section. Shi et al. [12] derived a two parameter micromechanics model to determine the effect of nanotube agglomeration on the elastic properties of randomly oriented CNTRC (Fig. 1). It is assumed that a number of CNTs are uniformly distributed throughout the matrix and that other CNTs appear in cluster form because of aggregation, as shown in Fig. 1. The total volume of the CNTs in the representative volume element (RVE), denote by $V_r$, can be divided into the following two parts:

$$V_r = V_r^{cluster} + V_r^m \tag{3}$$

The two parameters used to describe the aggregation are defined as:

IOP Conf. Series: Materials Science and Engineering **378** (2018) 012011 doi:10.1088/1757-899X/378/1/012011

$$\mu = \frac{Vcluster}{V}, \eta = \frac{V_r^{cluster}}{V_r} \quad 0 \le \eta, \mu \le 1 \tag{4}$$

where $V$ is volume of RVE, $V_{cluster}$ is volume of clusters in the RVE. $\mu$ is volume fraction of clusters with respect to the total volume $V$ of the RVE, $\eta$ is volume ratio of the CNTs inside the clusters over the total CNT inside the RVE.

When, $\mu = 1$ this means uniformly distribution of nanotubes throughout the entire composite without aggregation, and with the decreasing in $\mu$,the agglomeration degree of CNTs is more severe. When $\eta = 1$, all the nanotubes are located in the clusters. The case $\mu = \eta$ means that the volume fraction of CNTs inside the clusters is as same as that of CNTs outside the clusters so all CNTs are located randomly oriented. Thus, we consider the CNT-reinforced composite as a system consisting of clusters of sphere shape embedded in a matrix. We may first estimate respectively the effective elastic stiffness of the clusters and the matrix, and then calculate the overall property of the whole composite system. The effective bulk modulus K and shear modulus G of the cluster can be calculated by Prylutskyy et al. [13], respectively. The effective Young's modulus $E$ and Poisson's ratio $\upsilon$ of the composite can be calculated in the terms of K and G by Eq. (2).

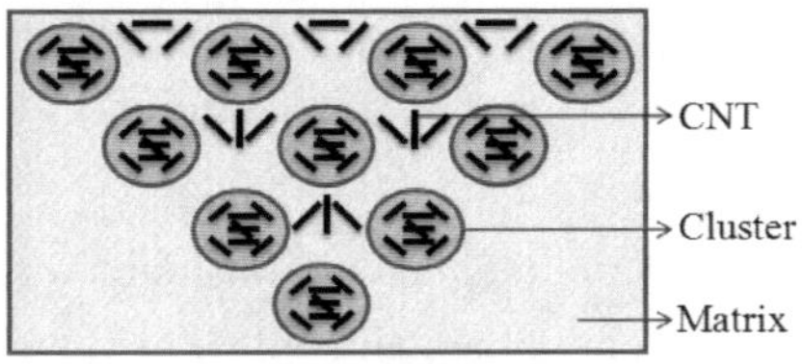


**Figure 1.** RVE with functionally graded Eshelby cluster model of aggregation of CNTs.

*2.3. CNT Volume Fraction Distribution*

Consider a rectangular FG-CNTRC plate with thickness of $h$, and the edges are parallel to axes $x$ and $y$, as shown in Fig. 2. The volume fractions of CNTs, $f_r$, are varied along the thickness of plate as the following:

$$f_r = (f_r^u - f_r^d)(1 + z/h)^p + f_r^d, -h/2 < z < h/2 \tag{5}$$

$$f_m = 1 - f_r \tag{6}$$

where $p$ $(0 \le p < \infty)$ is the volume fraction exponent and $f_r^u$ and $f_r^d$ are the values of CNT volume fraction at upper ($z$=$h$/2) and downer surfaces ($z$=-$h$/2), respectively. Fig. 3 shows the variation of CNT volume fraction along the thickness of plate for different values of the volume fraction exponent, $p$. Therefore the effective Young's modulus $E$ and Poisson's ratio $\upsilon$ are obtained from Eq. (2).

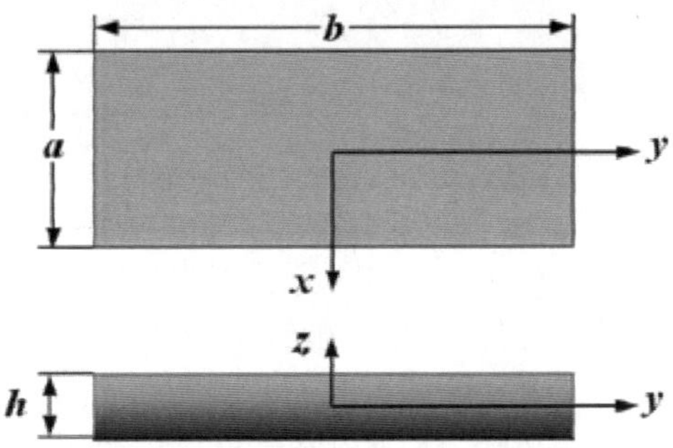


**Figure 2.** Schematic of the CNTRC plate.

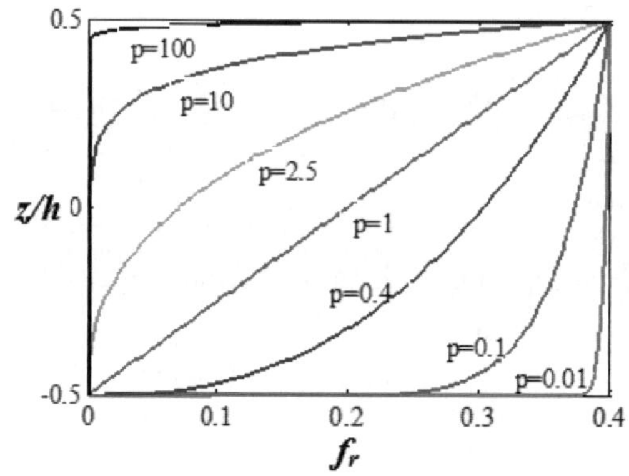


**Figure 3.** Variation of properties along the thickness of cylinders for different values of p according to Eq. (5)

*2.4. Governing Equations*

The mechanical constitutive relations that relate the stresses to the strains are as follows:

$$\sigma_{ij} = \lambda \varepsilon_{kk} \delta_{ij} + 2\mu \varepsilon_{ij} \tag{7}$$

where λ and μ are the Lame constants, $\varepsilon_{ij}$ is the infinitesimal strain tensor and $\delta_{ij}$ is the Kronecker delta. The infinitesimal strain tensor is related to the displacements as follows:

$$\begin{aligned} &\varepsilon_x = \frac{\partial u}{\partial x}, \varepsilon_y = \frac{\partial v}{\partial y}, \varepsilon_z = \frac{\partial w}{\partial z}, \\ &\gamma_{xy} = \frac{\partial u}{\partial y} + \frac{\partial v}{\partial x}, \gamma_{xz} = \frac{\partial u}{\partial z} + \frac{\partial w}{\partial x}, \gamma_{yz} = \frac{\partial v}{\partial z} + \frac{\partial w}{\partial y} \end{aligned} \tag{8}$$

where u, v and w, are displacement components along the x, y and z axes, respectively. Upon substitution (8) into (7) and then into the governing equations, the equations of motion are obtained in terms of displacement components. The related boundary conditions at z=-h/2 and h/2 are as follows:

$$\sigma_{zx} = \sigma_{zy} = \sigma_{zz} = 0 \tag{9}$$

Using the method of separation of variables, it is possible to seek solutions that are harmonic in time. The displacements can be written as follows:

$$\begin{aligned} &u = \sum_{m}^{\infty} \sum_{n}^{\infty} U(z) \cos(\beta_m x) \sin(p_n y) e^{i\omega t}, \\ &v = \sum_{m}^{\infty} \sum_{n}^{\infty} V(z) \sin(\beta_m x) \cos(p_n y) e^{i\omega t}, \\ &w = \sum_{m}^{\infty} \sum_{n}^{\infty} W(z) \sin(\beta_m x) \sin(p_n y) e^{i\omega t}, \\ &\beta_m = m\pi/a, p_n = n\pi/b, (m, n = 1, 2, \ldots) \end{aligned} \tag{10}$$

ω is the natural frequency and i(= $\sqrt{-1}$ ) is the imaginary number. Substituting for displacement components from (10) into the equations of motion which obtained in terms of displacement components, the coupled partial differential equations are reduced to a set of coupled ordinary differential equations (ODE). The geometrical and natural boundary can also be simplified, however, for brevity purpose they are not shown here. The differential quadrature method can be applied to discretize the equations of motion and the boundary conditions. One can find out how to use differential quadrature for discretizing the equations in [2-4] .The discretized form of the equations of motion and the related boundary conditions can be represented in the matrix form, and through solving an eigenvalue system of equations the natural frequencies and mode shapes of the plate can be calculated.

### 3. Numerical Results and Discussion

In this stage simply supported FG-CNTRC square plates are considered. In these plates the volume fraction of randomly oriented CNT, $f_r$, varies from zero to 0.4 according to Eq. (5) along the thickness of plate. Fig. 4 shows the first natural frequency parameters, $\Omega_{11}$, that are calculated by the following equation based on the mechanical properties of CNT for various values of $p$ and $b/h$.

$$\Omega = \omega b/h\sqrt{\rho^{CNT}/E^{CNT}} \tag{11}$$

This figure shows that increasing the ratio of width to thickness plates, $b/h$, or decreasing of the volume fraction exponent, $p$, can lead to increase $\Omega_{11}$. Fig. 5 illustrates variation of $\Omega_{11}$ versus μ for various values of η in UD-CNTRC square plate with a/h=10 and $f_r$=0.4. It can be seen that frequency parameter increases as the μ increases or especially η decreases. Also, Fig. 6 shows the first natural frequency of the same plates with $\eta$=1 and various values of $\mu$ and $a/h$. This figure shows that frequency parameters are increased by increasing of μ or decreasing of the ratio of length to thickness, $a/h$. Table 1 shows various modes of frequency parameters, $\Omega_{11}$, $\Omega_{12}$, $\Omega_{22}$, $\Omega_{13}$ for various values of p. it can be concluded that aggregation of CNTs deeply decreases frequency parameters in all modes. In the all of the above FG-CNTRC plates, volume fraction of CNT was changed but FG-CNTRC plates can be made by changing of the volume fraction of clusters. Consider CNTRC square plates with $a/h$=10, $f_r$ =0.2 and η=1. Volume fraction of clusters of CNTs, μ, varies from zero to 0.4 according to Eq. (5) along the thickness of plate. Table 2 shows various modes of frequency parameters, $\Omega_{11}$, $\Omega_{12}$, $\Omega_{22}$, $\Omega_{13}$, for various values of volume fraction exponent of clusters, p. This table discloses that increasing of p decreases frequency parameters in all modes. Finally, considers a UD-CNTRC square plate with $a/h$=10, $f_r$=0.2, η=1 and μ=0.5. Fig. 7 shows mode shapes of the plate at mode numbers of (1,1), (2,1), (1,2) and (3,1).

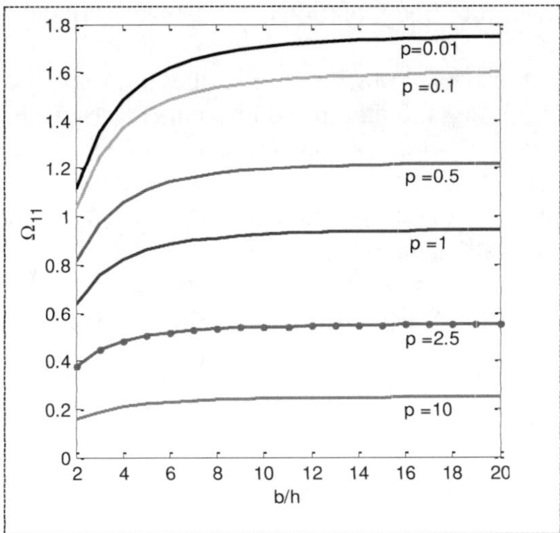


**Figure 4.** First frequency parameters versus b/h for FG-CNTRC square plates with fully dispersed CNT and $f_r$ =0→0.4.

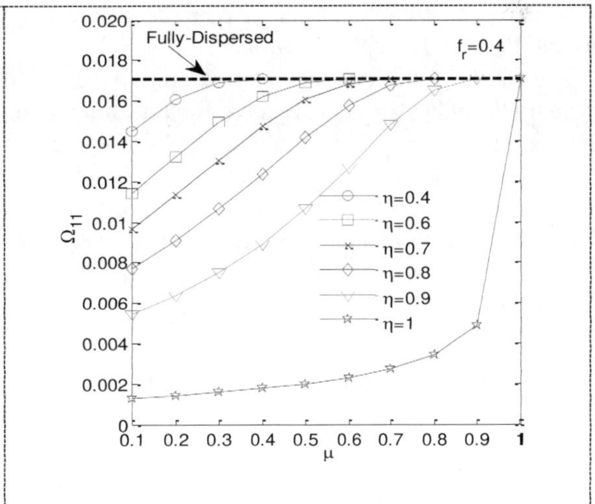


**Figure 5.** Variation of properties along the thickness of cylinders for different values of p according to Eq. (5)

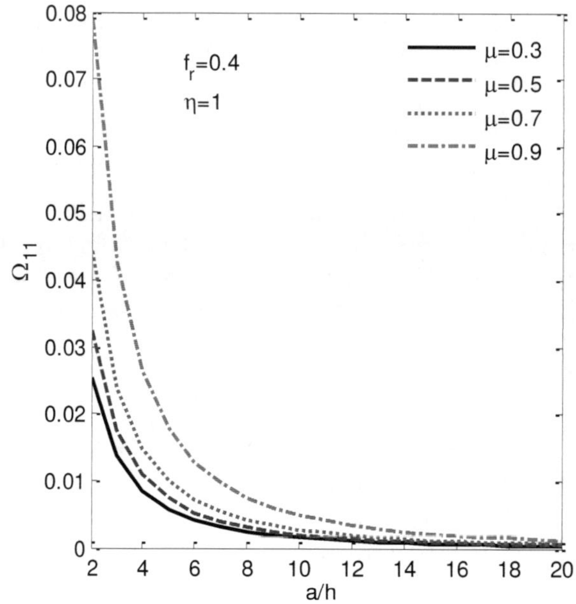


**Figure 6.** First frequency parameters versus *a/h* for UD-CNTRC square plates with aggregated CNT and $f_r$ =0.4.

## 4. Conclusion

In this research work, generalized differential quadrature method is employed to obtain a highly accurate semi-analytical solution for free vibration of functionally graded (FG) nanocomposite rectangular plates. The study is carried out based on the three-dimensional, linear and small strain elasticity theory. The Mori-Tanaka approach is implemented to estimate the effective material properties of the nanocomposite plate.

The agglomeration effect of single-walled carbon nanotubes, is considered in this study and it is shown that the natural frequencies of structure are seriously affected by the influence of CNTs agglomeration. Results presented the fact that mechanical properties and therefore free vibrations of plates are seriously affected by CNTs agglomeration.

The effects of various geometrical parameters, different material profiles along the thickness of rectangular plates are investigated. Results show that by increasing the ratio of width to thickness plates, *b/h*, or decreasing of the volume fraction exponent, *p*, the non-dimensional frequency of the plate increases. It can be concluded that aggregation of CNTs deeply decreases frequency parameters in all modes. It is observed that increasing of p decreases frequency parameters in all modes. It can be seen that frequency parameter increases as the μ increases or especially $\eta$ decreases.

Results reveal that natural frequency of the same plates with $\eta$=1 and various values of $\mu$ and *a/h*. It shows that frequency parameters are increased by increasing of μ or decreasing of the ratio of length to thickness, *a/h*.

**Table 1.** Frequency Parameters of FG-CNTRC Square Plates with a/h =10, μ=0.5, η=1 and $f_r$ =0→0.4.

| (m,n) | p=0.01 | p=0.1 | p=0.4 | p=1 | p=2.5 | p=10 | p=100 |
|---|---|---|---|---|---|---|---|
| (1,1) | 0.2002 | 0.2008 | 0.2016 | 0.1986 | 0.1774 | 0.1520 | 0.1327 |
| (1,2) | 0.4778 | 0.4790 | 0.4811 | 0.4743 | 0.4225 | 0.3612 | 03143 |
| (2,2) | 0.7332 | 0.7352 | 0.7385 | 0.7285 | 0.6559 | 0.5528 | 0.4796 |
| (1,3) | 0.8935 | 0.8959 | 0.9000 | 0.8882 | 0.8013 | 0.6725 | 0.5826 |

**Table 2.** Frequency Parameters of UD-CNTRC Square Plates with aggregated CNT, a/h=10, $f_r$ =0.2, $\eta$=1 and $\mu$=0→0.4.

| (m,n) | p=0.01 | p=0.1 | p=0.4 | p=1 | p=2.5 | p=10 | p=100 |
|---|---|---|---|---|---|---|---|
| (1,1) | 0.1791 | 0.1714 | 0.1569 | 0.1455 | 0.1376 | 0.1290 | 0.1203 |
| (1,2) | 0.4271 | 0.4091 | 0.3746 | 0.3468 | 0.3268 | 0.3058 | 0.2863 |
| (2,2) | 0.6554 | 0.6281 | 0.5753 | 0.5318 | 0.4995 | 0.4668 | 0.4385 |
| (1,3) | 0.7986 | 0.7655 | 0.7013 | 0.6478 | 0.6073 | 0.5672 | 0.5336 |

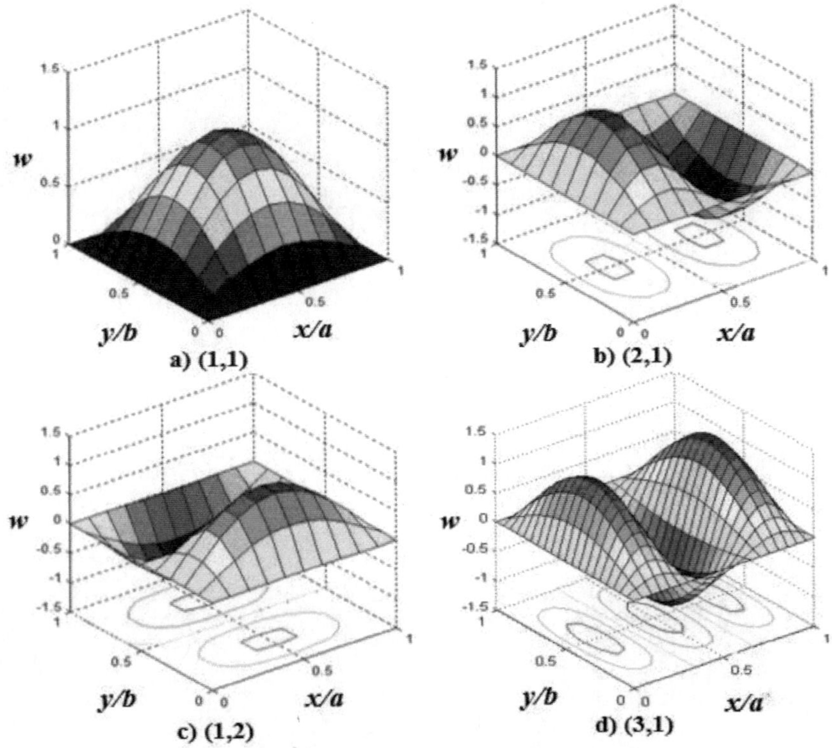


**Figure 7.** Different mode shapes of the plate.

## 5. References

[1] Wagner H D, Lourie O, and Y. Feldman 1997 *Appl. Phys. Lett.* **72** 188-190.

[2] Yas M H and Tahouneh V 2012 *Acta Mech.* **223** 43-62.

[3] Tahouneh V, Yas M H, Tourang H and Kabirian M 2013 *Meccanica* **48** 1313-1336.

[4] Tahouneh V 2016 *Steel and Composite Structures, An Internatinal Journal* **20** 623-649.

[5] Tahouneh V and Naei M H 2015 *J. Sandw. Struct. Mater.* **18**, 263-293.

[6] Tahouneh V and Naei M H 2013 *Meccanica* **49** 91-109.

[7] Cheng Z Q and Batra R C 2000 *J. Sound Vib.* **229** 879-895.

[8] Zenkour A M 2005 *Inter. J. Solid Struc.* **42** 5243-5258.

[9] Cheng Z Q and Batra R C 2000 *Archive Mech.* **52** 143-158.

[10] Reddy J N 2000 *Int. J. Numerical Methods Eng.* **47** 663-684.

IOP Conf. Series: Materials Science and Engineering **378** (2018) 012011 doi:10.1088/1757-899X/378/1/012011

[11] Matsunaga H 2008 *Compos. Struct.* **82** 499-512.

[12] Shi D L, Feng X Q, Yonggang Y H, Hwang K C and Gao H 2004 *J. Eng. Mater. and Technol.* **126** 250-257.

[13] Prylutskyy Y I, Durov S S, Ogloblya O V, Buzaneva E V and Scharff P 2000 *Comput. Mater. Sci.* **17** 352-355.

IOP Conf. Series: Materials Science and Engineering **378** (2018) 012012 doi:10.1088/1757-899X/378/1/012012

# Wear and Corrosion properties of CrSiCN coatings deposited by cathodic arc deposition process

**C H Yang, C Y Chen, and W Y HO**

Department of Materials and Energy Engineering, MingDao University, 369 Wen-Hua Rd., Petow, ChungHua, 523, Taiwan

E-mail: sampo0808@gmail.com

**Abstract.** In this study, three CrSiN based coatings were fabricated using cathodic arc deposition (CAD) technique. CrSiCN coatings were obtained by using metal plasma activated CAD process to decompose the $C_2H_2$ and $N_2$ reactive gas. The analyzed CrSiCN coatings were obtained with higher hardness and smooth surface. With addition of Si (8.7 at.% ) and C (41 at.%) in the coatings, the hardness increased to Hv3300 which is resulted from the addition of CrCN and CNx new phases. Tribological test shows the coefficient of friction of CrSiCN down to 0.2 was observed with the addition of Si and C, related to hardness and smooth surface. However, the corrosion resistance of the CrSiN coating is better than that of CrSiCN coating in the 1M $H_2SO_4$ solution at room temperature.

## 1. Introduction

Traditionally, CrN coatings are characterized by superior corrosion and oxidation resistance as compared to TiN coating. However, CrN coating is not as hard as TiN coating. By adding Si element into the CrN coating, CrSiN coatings have appeared better mechanical properties, tribological properties, and thermal stability [1-9]. CrSiN coatings, featured with nanocomposite structure that comprises nanocrystalline and amorphous phases, can be a potential material for tool applications. Previous studies have shown that the corrosion resistance of the CrSiN coatings was further improved by using cathodic arc deposition technology with pulsed current input [10]. The interfacial electrical resistance of CrSiN coatings deposited on AISI 304 stainless steel substrates similar with the blank substrate was confirmed. The improved corrosion behavior can be obtained by using the pulsed current of 90/120 and 90/150 A via cathodic arc deposition process [10]. Generally, the formation of nanocomposite structure and fine grain size can mainly contribute to the improved properties of the

IOP Conf. Series: Materials Science and Engineering **378** (2018) 012012 doi:10.1088/1757-899X/378/1/012012

CrSiN coatings. Carbon (C) is known to further reduce the friction coefficient and improve wear resistance of the coatings. Recently, CrSiCN coating was to be a potential coating as well [11-14]. The alloying effects from Si and C on the structure and properties of the CrSiCN coatings were to be investigated. The CrSiN or CrSiCN coatings were usually deposited via physical vapor deposition methods, especially magnetron sputtering techniques [14]. However, cathodic arc deposition (CAD) is known to be a vital technique to deposit hard coating on the cutting tools. In this study, CrSiN and CrSiCN coatings by using CAD process with different $C_2H_2$ and $N_2$ mixture as a reactive gas were compared in the field of the wear and corrosion properties.

## 2. Experimental

The CrSiN and CrSiCN coatings were deposited on AISI 304 stainless steel and tungsten carbide coupons in a cathodic arc deposition system (CAD). Two circular cathodic arc sources (dia. 150 mm) were installed face to face in the chamber, with a distance to substrate holder of 150 mm apart. One Cr target and one $Cr_{80}Si_{20}$ target were used as cathodic arc sources during the coating depositions. Prior to depositions, the chamber was pumped down to a base pressure below $1.0 \times 10^{-3}$ Pa. Then the substrates were cleaned by Ar+ glow discharged plasma at 1.3 Pa using a dc bias voltage of −800 V for 30 min. As for the coatings, layered structure containing Cr/CrN/CrSiN and Cr/CrN/CrSiN/CrSiCN was designed to insure the best adhesion with the substrate. Different ratio of $C_2H_2$ and $N_2$ mixture gas was introduced to form CrSiN and CrSiCN coatings via different flow rate. Scanning electron microscopy (SEM; model JOEL JSM-5600) was carried out to observe the surface morphology of the various samples. Chemical compositions of the coatings were measured using energy dispersive X-ray spectroscopy (EDS). A surface roughness measurement meter was used to measure the surface roughness of the coating. The hardness H was obtained using a Vicker's Hardness Tester under the different load of 0.01 to 0.05Kg. The tribological properties of the coatings were evaluated using a conventional ball-on-disc tribometer without any lubricant under a normal load of 5 N with a 6-mm diameter WC ball as a counterpart material. The corrosion resistance of the coated AISI304 stainless steel was investigated by in corrosive medium of 1M $H_2SO_4$ solution to simulate the aggressive aqueous environment. The electrode potential was scanned from −0.8 to 0.8 V at the scanning rate of 1.0 mV/s.

## 3. Results and Discussion

Fig. 1 shows the surface morphologies of the deposited CrSiN and CrSiCN coatings on WC-8%Co substrates with different $C_2H_2$ and $N_2$ mixture gas. It is clearly observable that the deposited CrSiN and CrSiCN coatings were dense structure, however, macroparticles emitted from the target were seen on the surface. Macroparticles are liquid or solid debris particles that are produced at cathode spots. Their presence is the most severe limitation to cathodic arc deposition technology. From the Fig. 1, the size of the macroparticles emitted from the metal targets was in the range up to 1μm which was related to the poison phenomena by introducing $C_2H_2$ and $N_2$ mixture gas in this study. Table 1 shows the surface roughness of the coatings. The $C_2H_2$ seems to reduce the size and quantity of the macroparticles. Therefore, the surface roughness of the CrSiCN coating (0.06μm) is smoother than that of CrSiN coating (0.11μm). The compositions of the coatings were analyzed by EDS as shown in Table 1. The higher $C_2H_2$ and $N_2$ mixture gas ratio leads to the higher C/N ratio in the coatings. However, Si content in the coatings exhibits a reverse trend of decreasing amount as the $C_2H_2$ gas increases. Previous study [10] shows the XRD patterns of the CrSiN coatings with (1 1 1), (2 0 0) orientation can be clearly indicated a fcc B1 NaCl-type crystal structure. As Si/(Cr+Si) concentration is about 18%, the CrSiN coating is exhibited a network structure which is comprises of CrN and SiNx phases [2]. Similar, the incorporation of Si and C in the coating may also induce phase transformation of crystalline fcc B1 NaCl-type structure (CrN, CrCN) and amorphous SiNx and CNx phase due to Si and C concentrations are high enough in this study.

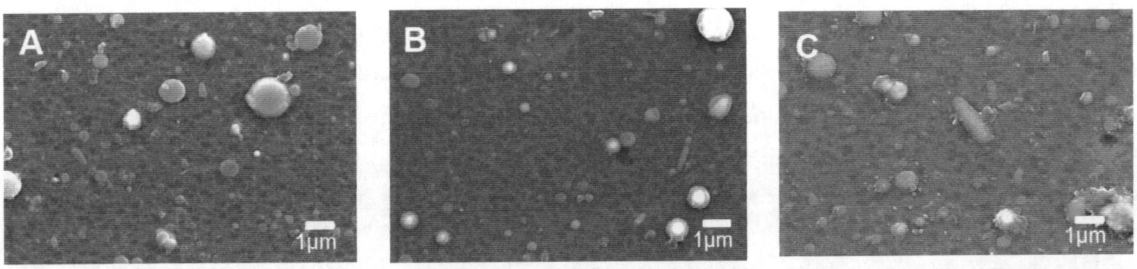


**Figure 1.** Surface morphologies of the coatings, A:CrSiN, B: CrSiCN, $C_2H_2$ and $N_2$ mixture gas ratio 1:1, C: CrSiCN, $C_2H_2$ and $N_2$ mixture gas ratio 3:1.

**Table 1.** Composition of the studied coatings analyzed by EDS

| sample | $C_2H_2/N_2$ ratio | Cr (at.%) | Si (at.%) | N (at.%) | C (at.%) | Roughness (Ra, μm) |
|---|---|---|---|---|---|---|
| A-CrSiN | 0/1 | 45.23 | 10.24 | 44.53 | -- | 0.11 |
| B-CrSiCN | 1/1 | 31.28 | 8.70 | 18.50 | 41.53 | 0.06 |
| C-CrSiCN | 3/1 | 33.43 | 5.18 | 7.93 | 53.45 | 0.06 |

Fig. 2 shows the hardness of the CrSiN and CrSiCN coatings obtained by Vicker's hardness measurement under different loading rate. The hardness of the samples, related to the increasing indenter depth with increasing load, is the combination of hardness of coating and substrate. The higher hardness is obtained by lowering the indenter load. The hardness of the CrSiN coatings was found to be Hv2850 at the load of 0.01Kg. Similar work was achieved that the highest hardness of the CrSiN coating was observed for the Si content around 7-10 at.% [9]. Both CrSiCN coatings increased its hardness with C introduced in the coatings. Apparently, the main effect can be anticipated by the solid solution hardening due to the substitution the Cr atom site for Si and C atom site. The hardness of the CrSiCN coating further increased to maximum values of Hv3300 (sample C) and Hv3000 (sample B), as the C increased to 41 and 53 at.%, respectively. Similar work by examining the XPS and TEM results [9, 14], a complete phase separation of CrN, CrCN nanocrystalline and SiNx and CNx amorphous phases has been achieved in CrSiN and CrSiCN coating, respectively. Therefore, it can be concluded that the hardness enhancement of CrSiCN coating is from the effects of solid solution hardening by CrCN nanocrystalline and the interface strengthening by SiNx and CNx.

Fig. 3 shows the coefficient of friction (COF) value of the CrSiN and CrSiCN coatings sliding against a WC–Co ball. In dry sliding tests, the CrSiN coating exhibited a similar COF of 0.5 to that of the previous work (COF=0.56) [9]. The COF value of CrSiCN coatings was decreased to 0.2 when the Si% was decreased below to 8.7 at.% and C% increased above to 41 at.%, respectively. Unlike CrSiN coating, the COF decreased obviously with the addition of Si and C, which can have a beneficial impact on the wear resistance. The decrease in the COF is mainly related to the increase of the hardness and smooth surface which is attributed to CrCN and CNx phases, acted as lubricant phase in the coatings. However, some peaks of COF during the wear test may be accompanied with crack of the coating due to the increase of the hardness. A better wear resistance of the CrSiCN coated tungsten carbide obtained is the potential application for the tools.

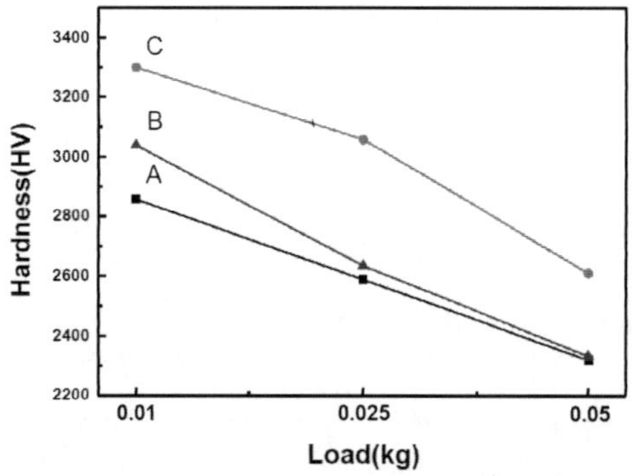


**Figure 2.** The hardness of the CrSiN and CrSiCN coatings obtained by Vicker's hardness measurement under different loading rate.

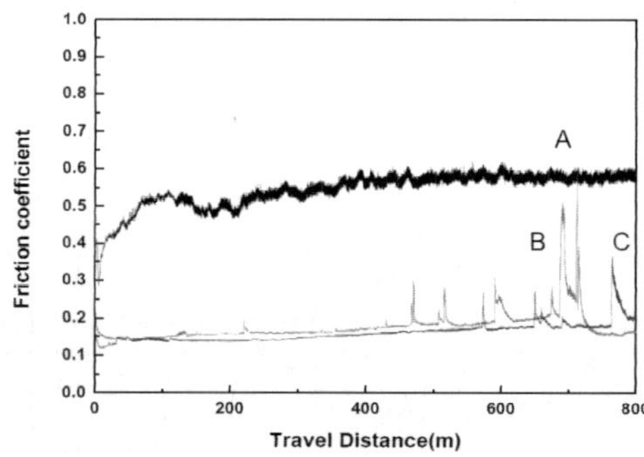


**Figure 3**. The coefficient of friction (COF) value e of the CrSiN and CrSiCN coatings sliding against a WC–Co ball.

Polarization tests were carried out using 1M $H_2SO_4$ solution at room temperature. The polarization curves (Ecorr. vs. Icorr.) of CrSiN and CrSiCN coated AISI 304 stainless steels are shown in Fig. 4. Compared with the coated specimens, the potentiodynamic polarization curves of CrSiN coating showed a more positive corrosion potential and a lower current density than those of the CrSiCN coated AISI 304 stainless steel substrate. Generally, Icorr commonly utilized as an index of corrosion rate is determined by the extrapolation of cathodic Tafel lines. In this study, a significantly reduced Icorr was observed for the CrSiN coating as compared to CrSiCN coatings. The CrSiN treated specimen had an evident enhancement on the corrosion resistance as compared to the CrSiCN coated specimens. With the complete phase separation of CrN nanocrystalline and SiNx amorphous phases, the CrSiN are well protected by the amorphous SiNx matrix. The strengthened interfaces can effectively limit the diffusion paths of the corrosive medium. For the CrSiCN coating, it is expected that higher amount of C element causes the less protection in the coatings. A better corrosion resistance of the CrSiN coated AISI 304 stainless steel obtained demonstrates that this coating can work as a protective barrier against the acid environment.

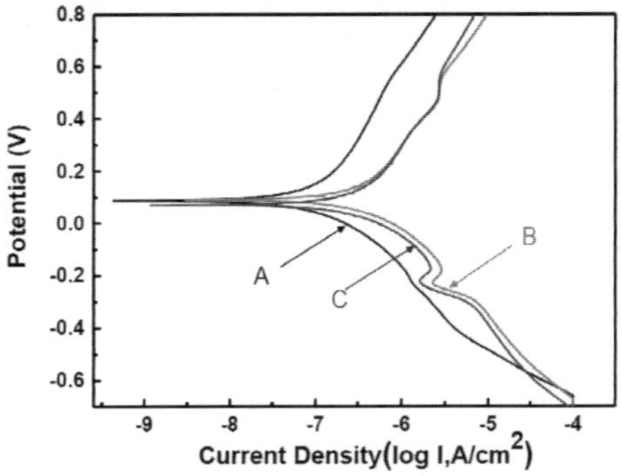


**Figure 4.** Polarization tests of CrSiN and CrSiCN coated AISI304 stainless steel carried out in 1M $H_2SO_4$ solution at room temperature.

## 4. Conclusions

In this study, three CrSiN based coatings were fabricated using cathodic arc deposition technique. By plasma decomposed $C_2H_2$ gas, additions of C to CrSiN were used to modify the microstructure and properties of the coatings. The analyzed CrSiCN coatings were obtained with higher hardness and smooth surface. With addition of Si (8.7 at.% ) and C (41 at.%) in the coatings, the hardness increased to HV3300 may result from the formation of CrCN and CNx new phases. Tribological property of the coefficient of friction down to 0.2 was observed with the addition of Si and C, related to hardness and smooth surface. However, the corrosion resistance of the CrSiN coating is better than that of CrSiCN coating in the 1M $H_2SO_4$ solution at room temperature.

## 5. Acknowledgement

Funding for this work was provided by the Ministry of Science and Technology, Taiwan, ROC under contract no. MOST 106-2221-E-451-001.

## 6. References

[1] Kim G S, Kim B S, Lee S Y, 2005 Surf. & Coatings Technol. 200 1814.

[2] Lee H Y, Jung W S, Han J G, Seo S M, Kim J H, Bae Y H, 2005 Surf. & Coatings Technol. 200 1026.

[3] Lee S Y, Hong Y S, 2007 Surf. & Coatings Technol. 202 1129.

[4] Mercs D, Briois P, Demange, Lamy V S, Coddet C, 2007 Surf. & Coatings Technol. 201 6970.

[5] Bousser E, Benkahoul M, Martinu L, Klemberg-Sapieha J E, 2008 Surf. & Coatings Technol. 203 776.

[6] Azzi M, Benkahoul M, Szpunar J A, Klemberg-Sapieha J E, Martinu L, 2009 Wear 267 882.

[7] Shah H N, Jayaganthan R, Kaur D, 2011 Applied Surf. Sci. 257 5535.

[8] Wo P C, Munroe P R, Li Z, Jiang Z T, Xie Z H, Zhou Z F, Li K Y, 2012 Mater. Sci. Eng. A 534 297.

[9] Lin J, Wang B, Ou Y, Sproul W D, Dahan I, Moore J J, 2013 Surf. & Coatings Technol. 216 251.

[10] Ho W Y, Tsou P Y, Chang Y S, Lin C L, 2014 Ad. Mater, Res. 960-961 152.

[11] Jeon J H, Jang C S, Yoon S Y, Shin B C, Kim K H, 2005 Surf. & Coatings Technol. 200 1635.

[12] Cai F, Huang X, Yang Q, Wei R, Nagy D, 2010 Surf. & Coatings Technol. 205 182.

[13] Wang Q Z, Wu Z W, Zhou F, Huang H, Niitsu K, Yan J, 2015 Surf. & Coatings Technol. 272 239.

[14] Lin J, Jang J, Park I W, Wei R, 2016 Surf. & Coatings Technol. 287 44.

IOP Conf. Series: Materials Science and Engineering **378** (2018) 012013 doi:10.1088/1757-899X/378/1/012013

# Hybrid Frequency-dependent Hysteresis Model of Magnetostrictive Actuator

**Yikun Yang, Bintang Yang and Muqing Niu**
State Key Laboratory of Mechanical System and Vibration, Shanghai Jiao Tong University, Shanghai, China 200240

E-mail: yyk199164@sjtu.edu.cn

**Abstract.** The inherent frequency-dependent hysteresis nonlinearities of magnetostrictive actuator prominently limit its performance. In this paper, a hybrid hysteresis model comprising a frequency-independent hysteresis model cascaded an adaptive finite impulse response filter is developed to improve the performance of actuators. For the frequency-independent part, a generalized Prandtl-Ishlinskii model that have analytical inversion is adopted. Besides, a modified coral reefs optimization algorithm is applied to identify the parameters to improve the model accuracy. The adaptive filter is utilized to adapt to the rate-dependent characteristic. The model validation results show that the hybrid model can satisfactorily describe the frequency-dependent hysteresis.

## 1. Introduction

Magnetostrictive actuators (MAs) are widely used in micropositioning, manufacturing applications and active vribration control systems due to their fast response and relatively large stroke and force capacity. Efficient precision positioning with magnetostrictive actuators is, however, limited by the intrinsic frequency-dependent hysteretic behaviour [1]. Inverse compensation method is the most popular approach to handle hysteresis by constructing a right inverse of the hysteresis model. So far, several models have been proposed to characterize the hysteresis phenomenon of MAs. In general, these models can be categorized into hysteresis models based on physic, such as Jiles-Atherton model and Duhem model, and hysteresis models based on phenomenology, such as the Preisach model and Prandtl-Ishlinskii (PI) model.

Phenomenological models can describe the behaviours of actuators well without physical mechanism, hence they can be convenient used in hysteresis compensation system. Among them, PI model with an analytical inverse model is the most commonly used. The classical PI model is a weighted superposition of several symmetric backlash operators, so it cannot describe the asymmetric hysteresis phenomenon, which is widely found in the MAs. To account for the asymmetry and saturation properties of hysteresis which are always observed for MA, Kuhnen [2] proposed a modified PI model with deadzone operators. Gu [3] introduced a generalized input function to describe the asymmetric hysteresis nonlinearity. Al Janaideh [4] proposed a generalized PI model with different envelope functions for increase curve and decrease curve, which can describe the symmetric and asymmetric hysteresis properties of MA with output saturation. However, the hysteresis nonlinearities of MA are dependent on the frequency of input, the above-mentioned models are frequency-independent.

To characterize the frequency-dependent hysteresis, several modified models, which formulate dynamic density functions in the classical models, have been developed. For instance, Al janaideh [5] proposed a PI model with dynamic density functions. As another choice, Al janaideh [6] modeled rate-dependent hysteresis of MA by PI model with dynamic thresholds. For all the rate-independent PI models, the analytical inverse can be obtained by the exact inversion law. However, great difficulties arise in the inversion calculation for the rate-dependent PI models that make those models difficult to compensate hysteresis in real-time applications. In some literature, the influence of the input frequency is due to the system dynamics of MA [7], thus the rate-dependent hysteresis nonlinearity of MA can be decomposed to a static hysteresis part and a linear dynamic part [8]. The adaptive filter can be utilized to model the rate-dependent characteristic of hysteresis by treat the frequency as an variable parameter. Besides, the adaptive filter can be identified online, that making the model can adapt the time-varying parameter. In this way, a rate-independent hysteresis model cascade an adaptive finite impulse response (FIR) filter is introduced to describe the rate-dependent hysteresis.

In this paper, the rate-dependent hysteresis is described by a cascaded model with an adaptive FIR filter and rate-independent hysteresis model with an generalized PI model that have analytical inversion. The hybrid model is discussed in section 2. Besides, a modified coral reefs optimization algorithm is utilized to identify the parameters to improve the model accuracy. The identification method and model validation are presented in section 3. Finally, conclusions are given in section 4.

## 2. Frequency-dependent hysteresis describing

As mentioned before, the hysteresis loops of the magnetostrictive actuator (MA) are frequency-dependent. In this work, a generalized Prandtl-Ishlinskii (GPI) model that have analytical inversion is applied as the rate-independent hysteresis model. Besides, an adaptive FIR filter model is introduced to series the GPI because it can adapt the variation of the frequency of input and conveniently implement the tracking algorithm. The structure of that hybrid model is shown in Fig. 1.

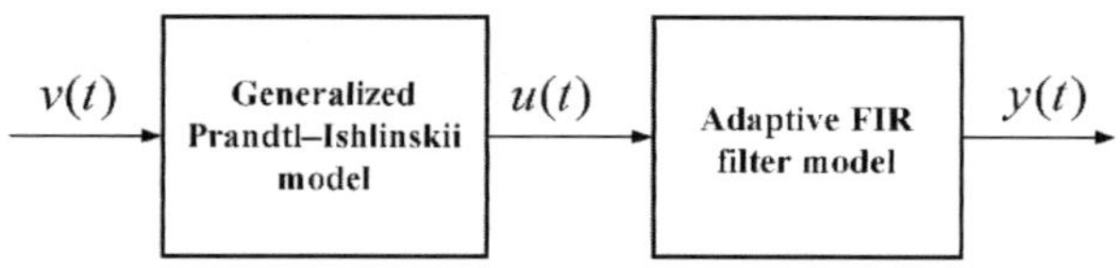


**Figure 1**. Frequency-dependent hysteresis model structure.

### *2.1. Frequency-independent GPI model*

The PI model describe the hysteresis input-output relationships based on play operator and density function. But the operator used in the classical PI model is symmetric, thus it cannot be applied for predicting asymmetric hysteresis of MAs. The generalized play operator is a nonlinear play operator resulting in asymmetric hysteresis loops, for the output increase and decrease, respectively, along different curves. The more details can be referred in [4].

A bias magnetic field is usually used to eliminate the frequency doubling effect of MA. Under a right bias field, the output displacement and input current of MA maintain positive. In this work, a one-side generalized play operator, as shown in Fig. 2, is adopted. The model based on this operator is a one-side generalized PI model (OGPI). The operator for input $v(t)$ and threshold $r$ is defined by [9]

$$G_{or}[v](t) = \begin{cases} g_{or}^{\zeta}(v(t), r, 0) & t = 0 \\ g_{or}^{\zeta}(v(t), r, G_{or}[v](t_i)) & 0 < t_i < t < T \end{cases} \quad (1)$$
$$g_{or}^{\zeta}(v(t), r, s) = \max(\zeta_l(v) - r, \min(\zeta_r(v), s))$$

where $\zeta_l$ and $\zeta_r$ are envelope functions to characterize the observed nonlinearities of MAs. The hyperbolic tangent envelope functions are chosen, since they can describe the output saturation properties of the actuators and have unique inverse function, which means the analytical inverse model can be obtained. The envelope functions are expressed as

$$\begin{aligned} \zeta_l &= a_0 \tanh(a_1 v + a_2) + a_3 \\ \zeta_r &= a_4 \tanh(a_5 v + a_6) + a_7 \end{aligned} \quad (2)$$

where all the coefficients which can be identified are constant and $a_0, a_1, a_4, a_5 > 0$.

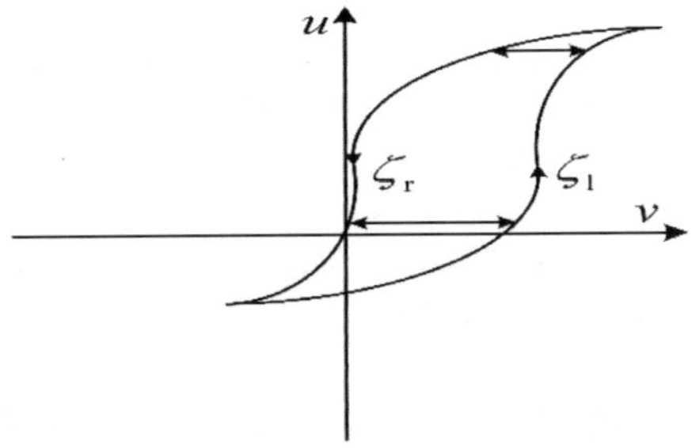


**Figure 2**. One-side Generalized play operator.

$\zeta_r$ always pass through the coordinate origin, therefore the coefficient should meet $a_4 \tanh(a_6)+a_7=0$. The finally output of the OGPI is

$$u(t)=\mathbf{w}^{\mathrm{T}}\ \mathbf{G}_{or}[v](t) \tag{3}$$

And the thresholds and weights: $r_i=\alpha i$ and $w_i=\beta e^{-\tau r_i}$, for $i=0,1,\cdots,n$. There $\alpha$, $\beta$ and $\tau$ are positive constants identified from experimental data of actuator.

### *2.2. Adaptive FIR filter*

In order to characterize the rate-dependent hysteresis, an adaptive FIR filter model is cascaded the OGPI model. An adaptive filter is a linear filter with transfer function controlled by variable parameters and those parameters is adjusted by an optimization algorithm. A delayed adaptive filter has been widely used for its simple structure and ease of implementation. Since the output generated by any given input is determined by these parameters, determining their values is the central task. The output of an adaptive tapped delay line FIR filter model is

$$y_k=\sum_{l=0}^{L}\omega_{lk}u_{k-l}=\boldsymbol{\omega}_k\mathbf{u}_k \tag{4}$$

where $L$ is the order of the filter, $\omega_{lk}$ is the $l$'th weight at $k$'th time. To get minimum error ($e_k$), $\boldsymbol{\omega}_k=\{\omega_{0k},\omega_{1k},\cdots,\omega_{Lk}\}$ needs to be updated by an optimization algorithm. In this way, adaptive FIR filter model can track the variation of the input frequency.

In concluded that adaptive FIR filter model is suitable for characterize the rate-dependent phenomenon of MA. Furthermore, FIR filter model can realize online modeling which endows the model with better adaptability. In conclusion, the hybrid model with an adaptive FIR filter cascade with a rate-independent OGPI model is introduced to describe the rate-dependent hysteresis. In this research, the order of FIR filter model is chose as $L=20$.

## 3. Model validation

### *3.1. Frequency-independent OGPI Parameter identification*

To identify the whole hybrid frequency-dependent hysteresis model, we identify the frequency-independent model at first. In OGPI model, there are 11 parameters to be identified

$$\mathbf{X}=\{a_o,a_1,a_2,a_3,a_4,a_5,a_6,a_7,\beta,\tau,\alpha\} \tag{5}$$

To identify the parameters of the frequency-independent model, a current at 0.1 Hz is applied. NI CompactRIO 9082 system is used to generate the control signal and acquired displacement signal. The control signal amplified by NF BP4610 is the input current of MA. A laser sensor (KEYENCE LK-G80) is used to measure the output displacement.

The degree of matching between target loop and simulated hysteresis loop is the judge standard of model accuracy. In this paper, the objective function is chosen as an error squared function between model output $u(i)$ and experimental displacement $s(i)$.

The coral reef optimization (CRO) algorithm is a novel global optimization algorithm, which mimics the behaviors of corals' reproduction and coral reef formation [10]. However, the local search capabilities of CRO is not well. In order to solve this problem, Rosenbrock's method (RM) [11] is introduced to fine-tune the solution around optimal.
First, a jump conditions is defined to determine whether to enter the RM process. In this work, the CRO process terminates and the RM process takes effect based on the best coral and the worst coral when $\Delta AF$ is bigger than the set threshold $S_{\text{break}} = 1\times10^{-6}$. The population size and maximum iterations are set as 50 and 300. The values of the step adjustment factors recommended by Rosenbrock are $\eta = 3.0$ and $\theta = -0.5$. The details about the MCRO can refer to our recent work [12].

### *3.2. OGPI Identification Results*

The identified **X** are list in Table 1. The identified hysteresis loops are shown in Fig. 3. For comparison, the identified loop using genetic algorithm (GA) is also plotted. Relative error (RE) is used to represent the accuracy:

$$\text{RE} = \sqrt{\sum_{i=1}^{N}(s(i)-u(i))^2 / \sum_{i=1}^{N}(s(i))^2} *100\% \tag{6}$$

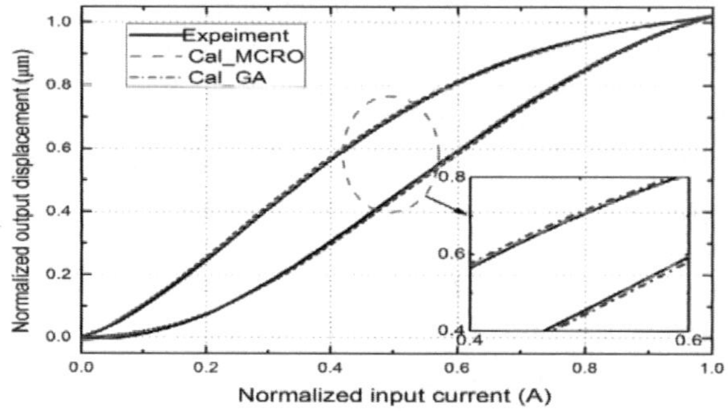


**Figure 3.** Identification results. Cal_MCRO: hysteresis loop identified by MCRO. Cal_GA: hysteresis loop identified by GA.

**Table 1.** Identified parameters of OGPI model.

| ***Parameters*** | ***Value*** |
|---|---|
| $\{a_0, a_1, a_2, a_3\}$ | 0.596, 1.665, -0.533, 0.317 |
| $\{a_4, a_5, a_6, a_7\}$ | 0.394, 2.056, -0.674, 0.223 |
| $\{\beta, \tau, \alpha\}$ | 0.196, 0.437, 0.056 |

The RE of loop identified by MCRO is 0.8167% is smaller than the corresponding RE of GA (1.023%). From the insert of Fig. 3, it can be observed that the loop identified by MCRO overlap the experiment loop better than the loop identified by GA.
The rate-dependent hysteresis loops of the used actuator are shown in Fig. 4. It can be observed that when the frequency increases the output displacement decreases and the loop becomes wider. Due to the bias field is introduced by the bias current, which will lead to more eddy current loss, the performance of actuator is already deteriorated at a relatively low frequency. As a consequence, the OGPI can satisfactorily describe the rate-independent hysteresis, but its performance will deteriorate when the input frequency changes.

### *3.3. Frequency-dependent Hybrid Model Identification*

The block diagram of the identification of the hybrid model is shown as Fig. 5. In this diagram, OGPI refers to the frequency-independent generalized PI model, $y_k$ is the output of the actuator, $\tilde{y}_k$ is the output of the identified FIR model and the optimization algorithm used to update the weights of FIR model is variable step-size normalized least mean square algorithm (VSNLMS) [13].
The weights update law can be expressed as

$$\boldsymbol{\omega}_{k+1} = \boldsymbol{\omega}_k + \mu_k \frac{\mathbf{u}_k}{\|\mathbf{u}_k\|^2} e_k \tag{7}$$

where $\mu_k = \mu_{max} \cdot \|\hat{\mathbf{P}}_k\|^2 / (\|\hat{\mathbf{P}}_k\|^2 + c)$ is the variable step size and $\hat{\mathbf{P}}_k = \gamma \hat{\mathbf{P}}_{k-1} + (1-\gamma)\mathbf{u}_k e_k / \|\mathbf{u}_k\|^2$. At the beginning, the size is large to get fast convergence speed and when the algorithm convergence the size is small to obtain small steady state error.

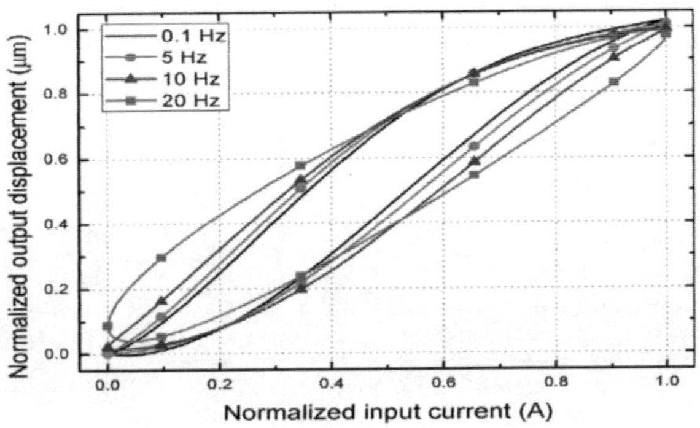


**Figure 4**. Frequency-dependent hysteresis loops.

An active vibration isolation used MA is used for experiment, so we mainly focused on the rate-dependent hysteresis model within a frequency range of 0.1 ~ 20 Hz. To validate the hybrid frequency-dependent hysteresis model for the MA, input current at a frequency of 0.1, 1, 10 and 20 Hz are applied to the MA, respectively. The parameters of weights update law are set as $\gamma = 0.95, c = 1, \mu_{max} = 0.12$. The accuracy of the model is measured in terms of root-mean-square error (RMSE) and relative error (RE) defined as

$$\text{RMSE} = \sqrt{\sum_{k=1}^{N} (\tilde{y}_k - y_k)^2 / N}, \ \text{RE} = \sqrt{\sum_{k=1}^{N} (\tilde{y}_k - y_k)^2 / \sum_{k=1}^{N} (\tilde{y}_k)^2} * 100\% \tag{8}$$

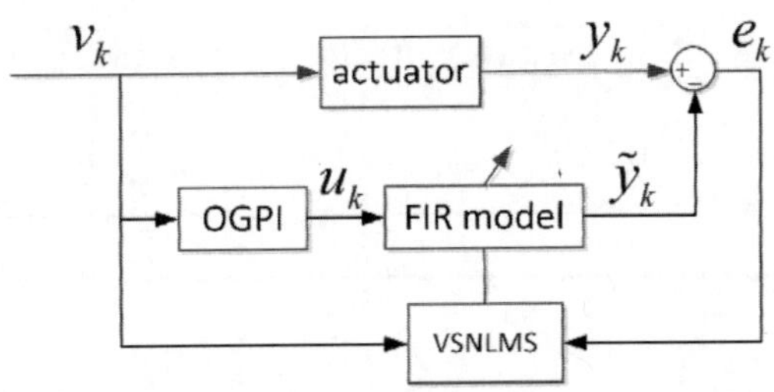


**Figure 5**. The block diagram of rate-dependent hybrid hysteresis model identification.

The experimental hysteresis loops and loops simulated based the rate-dependent hybrid hysteresis model (RDHM) are shown in Fig. 6. As a comparison, the simulated loops of generalized PI model alone are also depicted. The accuracies of those models are listed in Table. 2. It can be observed that the accuracy of OGPI model is unsatisfactory, but the proposed model can describe the frequency-dependent hysteresis fairly accurately from 0.1 Hz (super-low frequency) to 20 Hz. The RMSE of the hybrid model are less than 1 μm. The RE of the hybrid model is 1.64% when the frequency of inputs is 20 Hz, it is less than 1% under other frequencies. Moreover, the FIR filter part of the proposed model is build online, so the hybrid model can track the variation of system. Compared to the frequency-dependent PI model [5] and [6], the proposed hybrid hysteresis model is easy to implement in the hysteresis compensation control system.

## 4. Conclusions

In this paper, we proposes a hybrid frequency-dependent hysteresis model for magnetostrictive actuator. This hybrid hysteresis model comprise a generalized Prandtl-Ishlinskii static hysteresis model that has analytical inversion and an adaptive FIR filter. A one-side operator is utilized in this model because of the output displacement and input current of MA are always positive. Besides, a modified coral reefs optimization algorithm is applied to identify the parameters to improve the accuracy. The developed model can describe the hysteresis loops well, the root-mean-square error of the proposed model are less than 1 μm. In the future work, the proposed model should be applied in the hysteresis compensation control system for magnetostrictive actuators and other smart material based actuators.

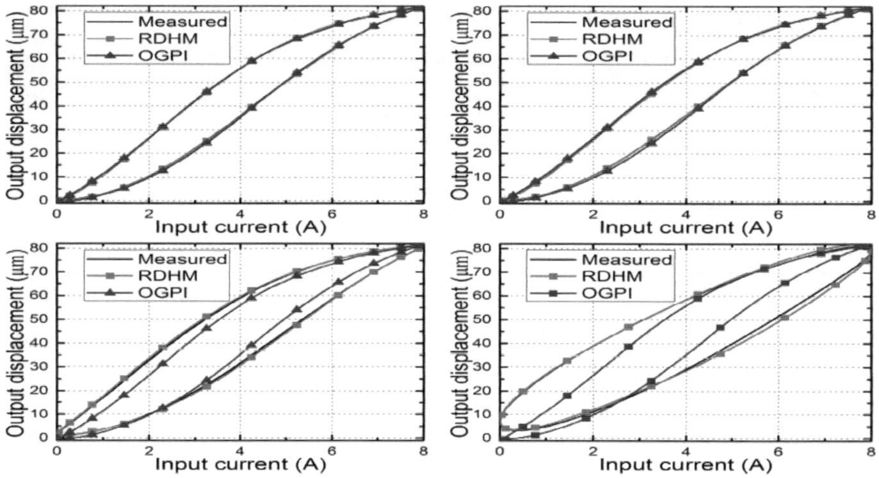


**Figure 6.** Model validation in view of frequency change. (a) 0.1 Hz. (b) 1 Hz. (c) 10 Hz. (d) 20 Hz. RDHM: the proposed rate-dependent hysteresis model, OGPI: one-side generalized PI model alone.

**Table 2.** Accuracy of model

| *Frequency* | *RDHM* | | *OGPI* | |
|---|---|---|---|---|
| *(Hz)* | RMSE(μm) | RE(%) | RMSE(μm) | RE(%) |
| 0.1 | 0.074 | 0.142 | 0.478 | 0.876 |
| 1 | 0.116 | 0.219 | 0.650 | 1.20 |
| 10 | 0.442 | 0.825 | 3.24 | 6.05 |
| 20 | 0.991 | 1.89 | 7.71 | 14.7 |

## 5. Acknowledgments
The work is supported by Shanghai Jiao Tong University and China Aerospace Science and Technology Group (SJTUCASC) Advanced Space Technology Fund USCAST2016-13 and the research grant (1600158) from Qinghai Province Key Laboratory of Photovoltaic grid connected power generation technology, for which the authors are most grateful.

## 6. References
[1] X. Tan, J.S. Baras 2004 *Automatica* **40(9)** 1469-1480.
[2] K. Kuhnen, Modeling 2003 *Eur J Control* **9** 407-418.
[3] G.-Y. Gu, L.-M. Zhu, C.-Y. Su 2014 *IEEE T Ind Electron* **61** 1583-1595.
[4] M. Al Janaideh, S. Rakheja, C.-Y. Su 2011 *IEEE-ASME T Mech* **16** 734-744.
[5] O. Aljanaideh, M.A. Janaideh, S. Rakheja, C.-Y. Su 2013 *Smart Mater Struct* **22** 025027.
[6] O. Aljanaideh, S. Rakheja, C.-Y. Su 2014 *Smart Mater Struct* **23** 035002.
[7] Z. Li, X. Zhang, G.Y. Gu, X. Chen, C.Y. Su 2016 *IEEE T Ind Inform* **12** 980-990.
[8] E.V. Papoulis, T. Stathaki 2004 *IEEE Signal Proc Let* **11** 56-59.
[9] Y. Yang, B. Yang, M. Niu 2018 *Nonlinear Dynam* **9** 1-10.
[10] S. Salcedosanz, S.J. Del, I. Landatorres, S. Gillópez 2014 *Thescientificworldjo* **2014** 739768.
[11] F. Kang, J. Li, Z. Ma 2011 *Information Sciences* **181** 3508-3531.

[12] Y. Yang, B. Yang, M. Niu 2017 *Sensor Actuat A-Phys* **261** 184-195.
[13] H.C. Shin, A.H. Sayed, W.J. Song 2004 *IEEE Signal Proc Let* **11** 132-135.

IOP Conf. Series: Materials Science and Engineering **378** (2018) 012014 doi:10.1088/1757-899X/378/1/012014

# Study on Oxidation Resistance of TiAlN Films Deposited by Magnetron Sputtering

**Hongying HUO[1] and Min ZOU[2]**

[1]State Key Laboratory for Vanadium & Titanium Testing, Panzhihua University, Panzhihua, China

[2]College of Material Engineering, Panzhihua University, Panzhihua, China

E-mail: 258116574@qq.com

**Abstract.** In this paper, TiAlN hard thin film is deposited on hard alloy surface by magnetron sputtering; at 400 °C to 825 °C, oxidation experiment is performed for the samples; XRD is used for phase analysis of TiAlN hard thin film; Through the use of HVS-1000 digital micro-hardness tester and WS-2004 automatic scratching tester, microhardness of the samples and substrate-film adherence are analyzed. Results show: the microhardness and coherence of TiAlN thin film reduces with the rise of temperature; at 800 °C, the microhardness and substrate-film adherence remain at 1603 HV and 48 N respectively; it is found through comprehensive judgment that TiAlN thin film can work under 800 °C.

## 1. Introduction

The main forms of failure of parts in the working process are wear, corrosion and fatigue. Especially in high temperature, high speed, high pressure, working conditions, such as overloading, corrosion, parts often due to surface damage and the local failure, lead to equipment instability or even scrapped. Therefore, the use of various techniques to improve the surface performance of the parts is to increase the safety and reliability of the parts, and to prolong the service life. Surface membrane technology is in parts coated on the surface of a particular material to achieve important technology to improve the surface performance of the parts, the parts surface with very little material of high hardness, high temperature resistant, corrosion resistant, and special properties such as light, heat, electricity, magnetism, a need to use a large number of whole material to achieve or to meet the performance, thus get rapid development in recent years, has been widely used in machinery manufacturing, aerospace, optical engineering, computer engineering, electronic technology [1-2].

In modern industry, cutting tools are required to operate at higher and higher speed. And the frictional heat causes cutting blades to be in high temperature. Therefore, there are higher requirements on the red hardness, wearability, thermal stability and antioxidation of cutting tools. For its hardness, wearability and high oxidation resistance, TiAlN thin films are adopted by hard alloy cutters, so as to significantly promote cutter performance and service life [3-4]. There have been studies on high temperature oxidation property of TiAlN, which mostly focus on the influences of preparation methods, process and additional elements [5-7]. There have been no reports on the effects of oxidizing temperature on film hardness and binding force. In this paper, unbalanced magnetron sputtering is used for preparation of TiAlN thin film on the substrate of hard alloy; isothermal oxidation experiment is performed in different temperature, so as to analyze the effects of oxidation temperature on microhardness and substrate-film adherence.

## 2. Experiment

Hard alloy numbered with YG6 is used as the substrate in this experiment. To ensure the cleanness of the substrate surface, substrate needs pretreatment: clamping the substrate; add detergent solution into

IOP Conf. Series: Materials Science and Engineering **378** (2018) 012014 doi:10.1088/1757-899X/378/1/012014

ultrasonic machine, and heat the solution to about 60 °C; put the substrate in the ultrasonic machine and clean it for 10 minutes; dry the substrate with high pressure nitrogen and then put it into baker for 10 minutes; start the experiment when the substrate is completely dry.

Main instrument: JGP450 multi-functional magnetron sputtering machine. The sputtering target is made of titanium aluminum alloy— Ti:Al=7:3; the working gas is argon (with purity above 99.999%), and the reacting gas is nitrogen (with purity higher than 99.999%). Parameters pertaining to deposition process are as follows— base pressure: 5×10-3Pa; deposition pressure: 4×10-1Pa; deposition current:120 A; substrate bias voltage: -120 V; partial pressure of Ar: 1.3E-1 Pa; nitrogen flow: 60mL/min; sputtering time: 120 min.

The isothermal oxidation experiment is performed in electric resistance furnace with static air, where the heating temperatures are set as 400 °C, 500 °C, 600 °C, 700 °C, 800 °C and 825 °C.、Each time following the heating processes, which are numbered with 2 to 7, there is heat preservation of one hour; the unheated sample is numbered with 1.

X-ray diffractometer (XRD) is used for analysis of the samples; HVS-1000 digital micro-hardness tester is utilized to t-st the sample hardness; WS-2004 automatic scratching tester is employed to test the substrate-film adherence.

## 3. Results and analyses

3.1 Results and analysis of TiAlN film oxidation

The quality before and after oxidation the sample of 2~7 was measured respectively, which were measured by the FA1104 electronic precision scales ,produced by Shanghai balance factory, and accuracy was 0.1mg . Sample number and oxidation parameters are shown in Table 1. The experimental results are shown in Table 2.

The table 2 shows ,the samples were not nodal increment at 400 °C and 500 °C and insulation 30 min. even aerobic element diffusion under 500 °C, the amount is very small, under one over ten thousand g orders of magnitude. the sample is 0.2 mg of weight gain at 600 °C, weight gain is not obvious, so oxidation resistance of film is still very good. The quality of the sample was larger increase( 4.4 mg) at 700 °CCompared to at 600 °C.When the oxidation temperature up to 800 °C, oxidation is more serious, sample weight to 23 mg. When the oxidation temperature of 825 °C, the sample weight is 30.9 mg. Temperature will be above 600 °C oxide with quadratic polynomial fitting quality weight, quality weight gain and the relation between the oxidation temperature:

$$Y=0.0008X^2-0.9488X+295.75 \quad (3.1)$$

Correlation coefficients R2 = 0.9995.

The formula for the rate of sample oxidation of the 2.1 derivation:

$$Y'=0.0016X-0.9488 \quad (3.2)$$

As the oxidation temperature increases, the oxidation rate also increases.

**Table 1** The technology parameters of oxidation experiment

| Number | Oxidation temperature /°C | Holding time /min | Cooling way |
|---|---|---|---|
| 1# | Room temperature | — | furnace cooling |
| 2# | 400 | 30 | furnace cooling |
| 3# | 500 | 30 | furnace cooling |
| 4# | 600 | 30 | furnace cooling |

IOP Conf. Series: Materials Science and Engineering **378** (2018) 012014 doi:10.1088/1757-899X/378/1/012014

| | | | |
|---|---|---|---|
| 5# | 700 | 30 | furnace cooling |
| 6# | 800 | 30 | furnace cooling |
| 7# | 825 | 30 | furnace cooling |

**Table 2** The results of oxidation experiment

| Number | 1# | 2# | 3# | 4# | 5# | 6# | 7# |
|---|---|---|---|---|---|---|---|
| Before oxidation /g | 5.1677 | 5.2179 | 5.1567 | 5.1434 | 5.1434 | 5.1707 | 5.1628 |
| After oxidation /g | 5.1677 | 5.2179 | 5.1567 | 5.1436 | 5.1436 | 5.1937 | 5.1937 |
| Weight increment /mg | 0 | 0 | 0 | 0.2 | 4.4 | 23 | 30.9 |

3.2 TiAlN thin-film phase analysis

Figure 1 represents the XRD spectrogram for film preparation, which shows the two phases of the samples: WC and $Ti_3AlN$— the former is the major constituent of the hard alloy and the latter the component of $Ti_3AlN$. The spectrogram shows no other phases. It's known that there are 4[8] Ti-Al-N phases: $Ti_{0.5}Al_{0.5}N$, $Ti_2AlN$, $Ti_3AlN$ and $Ti_{1.5}AlN$, among which $Ti_3AlN$ is a common phase. From figure 2-5, the oxidation temperature is below 600 °C, there is no change in the phase composition of the sample. $Ti_6O_{11}$ from 600 °C to a new phase, the phase at 700 °C still exist, but contrast method was 600 °C, 700 °C $Ti_6O_{11}$ phase characteristic peak of map has been obviously reduced, and at 800 °C $Ti_6O_{11}$ disappear. The literature[9] considers that the oxidation of TiAlN films is divided into two processes.The first is the diffusion process of O atoms to the inside of the film, while Ti atom and Al atom are diffused at low temperature.With oxidation temperature rise, Al, Ti atoms to the film surface diffusion, by contrast, Al atoms to the film surface diffusion speed is higher than that of Ti atom diffusion velocity, thus rich Al poor Ti in film surface, accordingly, the inner rich Ti poor Al. Therefore, first of all, on the surface of the membrane, the Al atom combined with O atoms to form the oxide layered $Al_2O_3$, and a small amount of oxygen formed $TiO_x$ by reaction with Ti rthrough the $Al_2O_3$ oxide layer , forming the $Al_2O_3$ / $TiO_x$ two oxide layered structure.

At above 800 °C, O(oxygen) begins to reacts with the hard alloy substrate, generating $CoWO_4$. This indicates that 1TiAlN thin film has been seriously damaged, and there is weakened or no separation between external oxygen and the substrate. Meanwhile, it can be seen from Figure 6 that the characteristic peak pertaining to $Ti_3AlN$ is still noticeable, which indicates that the thin film has not been totally oxidized or dropped— it is still functioning.

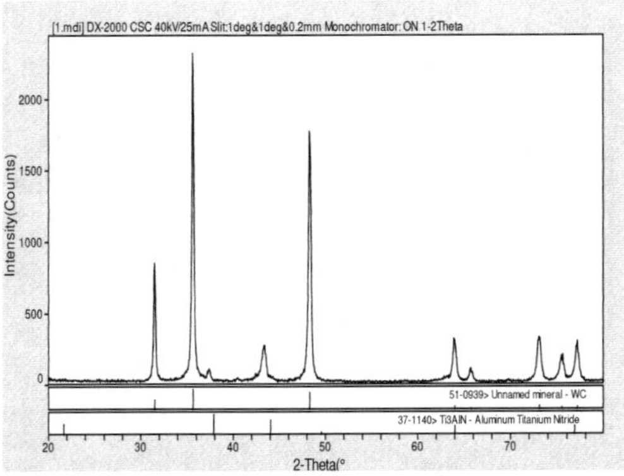


**Figure 1** The X-ray diffraction pattern of No.1

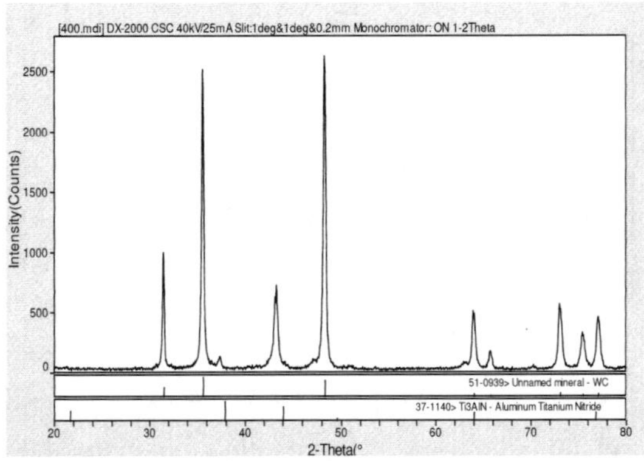


**Figure 2** The X-ray diffraction pattern of No.2

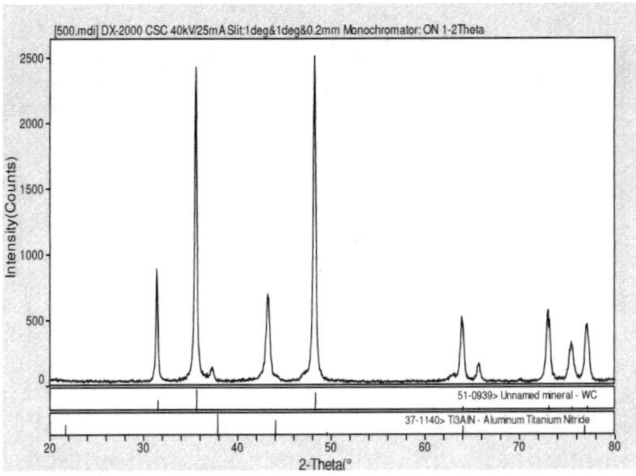


**Figure 3** The X-ray diffraction pattern of No.3

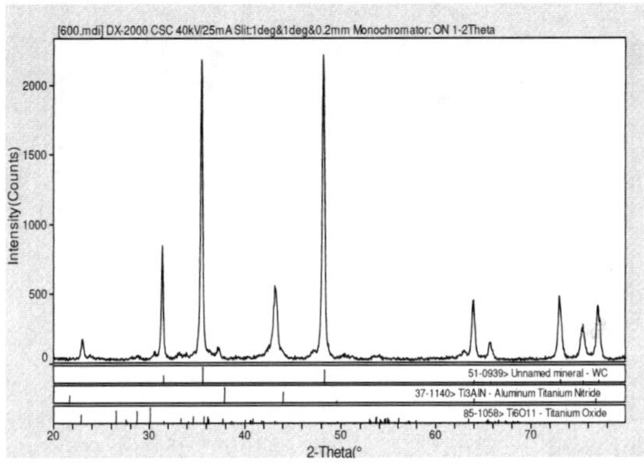


**Figure 4** The X-ray diffraction pattern of No.4

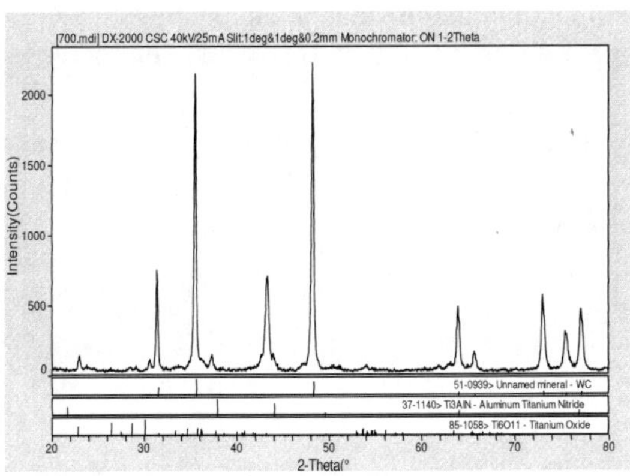


**Figure 5** The X-ray diffraction pattern of No.5

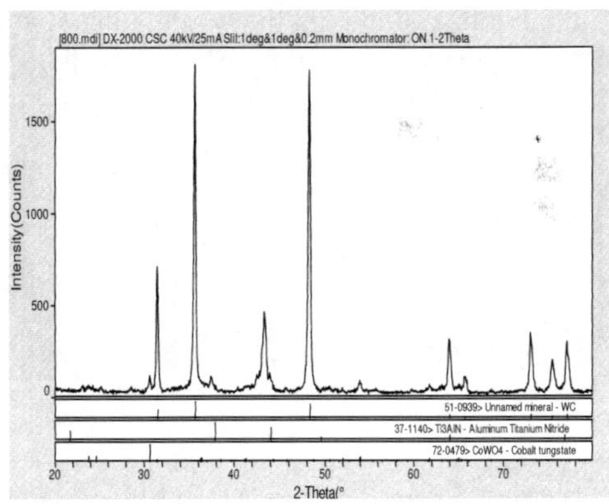


**Figure 6** The X-ray diffraction pattern of No.6

3.3 Analysis of the effects of oxidation temperature on TiAlN film hardness

In this experiment, HVS-1000 digital micro-hardness tester is adopted, with the indenter loaded with 100 gf[10]. The test results are shown in Table 3. It can be seen from the table that the film hardness undergoes no significant changes below 400 °C. This is because the $TiO_2$ generated at low temperature is not enough to have effects on the overall hardness. With the oxidation temperature rising, the film hardness decreases obviously; when the oxidation temperature reaches 825°C, the mircohardness of the sample is close that of hard alloy substrate. The oxidation causes the reduction in film densification; the phase changes and increased stress in the film, etc. can lead to reduction of film microhardness.

Many data show that the microhardness of TiAlN film in more than 3000 hv. However, because of the thin film thickness low, .this paper's microhardness of thin film is low. For a thin film with very low thickness, the hardness of the film is difficult to avoid without the influence of the substrate, and the microhardness value measured is the compound hardness of the film and the matrix. The thickness of the film thickness, the smaller the influence of the substrate, the closer the hardness of the measured microhardness is. On the contrary, the thickness of the film is thinner, and the larger the influence of the substrate, the closer the measured microhardness is to the hardness of the substrate. The thinness of the TiAlN film is the cause of low hardness.

**Table 3** The micro-hardness of these samples in different temperature

| Temperature/°C | Room temperature | 400 | 500 | 600 | 700 | 800 | 825 |
|---|---|---|---|---|---|---|---|
| Micro-hardness /Hv | 2580.0 | 2168.3 | 1778.6 | 1752.6 | 1672.2 | 1603.0 | 1509.0 |

3.4 Analysis of the effects of oxidation temperature on TiAlN film adherence
Substrate-film adherence is a very important film indicator. Currently, scratching is mostly used for test of film adherence[11]. In this experiment, WS--2004 automatic scratching tester is utilized for test of substrate-film adherence in the samples. Table 4 demonstrates the adherence under different oxidation temperature. It can be seen from it that oxidation temperature has significant effects on the adherence. Below 800°C, the adherence declines gently— this is due to the differences between thermal expansion coefficients of the substrate and thin film. Accordingly, the stress between them increases, which results in the deduction of substrate-film adherence. Above 800°C, the adherence undergoes dramatic changes, when thermal expansion coefficients are no longer the major factors causing adherence reduction; with oxidization temperature rising, new phases emerge in the film, changing the film's thermal expansion coefficient, which has effects on the substrate-film adherence. Through XRD analysis, it can be seen that the composition of thin film has changed significantly; large amounts of $TiO_2$ have been generated under the temperatures; the reduction of TiN and increase of $TiO_2$ are the major factors in adherence reduction.

**Table 4** The adhesion strength in different temperature

| Temperature/°C | Room temperature | 400 | 500 | 600 | 700 | 800 | 825 |
|---|---|---|---|---|---|---|---|
| Adhesion strength /N | 64 | 58 | 57 | 55 | 50 | 48 | 38 |

## 4. Conclusion

(1). TiAlN thin film has been prepared using unbalanced magnetron sputtering; the film hardness reaches 2580 HV, and the coherence 64 N.

(2). The microhardness and coherence of TiAlN thin film reduces with the rise of temperature; at 800 °C, the microhardness and substrate-film adherence remain at 1603 HV and 48 N respectively; it can be found through comprehensive judgment that TiAlN thin film can work under 800 °C.

## 5. REFERENCES

[1] Qian Miaoshan, Yao Shoushan, Zhang Shaozong. Modern surface technology [M]. Beijing: mechanical industry press, 2000.
[2] Qu Jingxin, Wang Honghong. Surface engineering handbook [M]. Beijing: chemical industry press, 1998.
[3] Paul H M,Christian M. Microstructrual design of hard coatings[J].Progress in Materials science. 2006(51):1080-1110.

[4] Ma L M,Cairney J M,Hoffman M J.Surface and Coating Technology,2006,200:3518
[5] Kang Bo, Ma Ruixin, Wu Zhongliang, et al., Research Status of Modern Cutter Film Preparation Technology[J], Surfaces Technology, 2008, 38(2):71-74.
[6] Yan Liangchen, Xiong Xiaotao, Yang Huisheng, et al., (Ti, Al)N Films Deposited by Double-Target Reactive Magnetron Co-Sputtering[J], Vacuum Science and Technology, 2005, 25(3):233-237.
[7] Li Zhi, Cui Jingzhong, Li Guanbin, et al. Research on TixAlyNz Thermal Control Films Deposited by Reactive Magnetron Sputtering[J], Vacuum and Cryogenics, 2005, 11(1):34-39.
[8] M.Zlatanovic.Surface and Coatings Technology.1990,41:17
[9] Wang Yongkang, Lei Tingquan, Xia lifang et al., the high temperature oxidation behavior of Ti0.5 Al0.5 N coating [J]. Materials engineering, 2001 (1) : 12-14
[10] Lai Qianxi, Tian Jiawan, Dai Jiawei, et al., Load Selection in Microhardness Measurement of Hard Thin Films, Journal of Chinese Electron Microscopy Society, 2002, 21(5): 635-636
[11] Qu Quanyan, Qiu Wanqi, Zeng Dechang, et al., Comprehensive Evaluation of Substrate-Film Adherence by Scratching[J], Vacuum Science and Technology, 2009,29(2):184-187

IOP Conf. Series: Materials Science and Engineering **378** (2018) 012015 doi:10.1088/1757-899X/378/1/012015

# Deformation Characteristics and Texture Development of Bi-added Cu-29Zn Alloys

**Imam Basori[1,2], Irene Angela[1], David Jendra[1], and Bondan T. Sofyan[1]**

[1]Department of Metallurgy and Materials Engineering, Faculty of Engineering, Universitas Indonesia, Kampus UI Depok 16424, Indonesia

[2]Department of Mechanical Engineering, Faculty of Engineering, Universitas Negeri Jakarta, Jl. Rawamangun Muka, Jakarta 13220, Indonesia

Corresponding author: bondan@eng.ui.ac.id

**Abstract.** Theories regarding texture transition mechanisms of brass alloys and its critical deformation level of when the change of texture transition occurs are still in debate. Previous observation showed that copper type texture dominantly formed at high temperature process whereas brass type texture tended to occur at low temperature. In contrast, another research observed the occurrence of copper type texture at low strain level while higher strain level resulted in brass type texture. Thus further research is needed to affirm one that well-satisfy the actual texture transition phenomenon. In this research, Cu-Zn-xBi alloys were produced by gravity casting process using pure Cu and Zn ingots with varied Bi addition of 0.2, 0.4, and 0.8 wt. %. As-cast samples were homogenized at 800 °C for 2 h in a muffle furnace. The samples were then cold-rolled with the level of deformation of 20, 40, and 70 %. Characterization includes chemical composition analysis, microstructure observation, hardness testing, and texture measurement. Results showed that addition of Bi prompted Bi-rich dispersoid phase which segregated inside the grain and along the grain boundary in globular forms. Slip was dominant at 20 % deformation level and its density increased with the addition of Bi. At 40 % deformation, twinning replaced slip as the predominant mechanism. The twin density increased with higher Bi content. Further deformation at 70% produced shear bands and flattened the Bi dispersoid phase. Greater Bi content induced the formation of shear band. At 40% deformation, pole figure images show the trend of copper type texture with orientation of {112} <111> with intensity of 7.92. This texture was formed by mechanical twinning during deformation process. At 70 % deformation, samples illustrated the combination of brass and Goss type texture at orientation of {110} <112> and {110} <001> due to the shear band.

Keywords: Cu-29Zn, Deformation, Cold rolling, Slip, Twinning, Shear band, texture.

## 1. Introduction

Metal forming process has to be accompanied by good formability of material, particularly those done in high deformation level. Formability is highly influenced by substantial factors such as solid crystal type, grain size, alloy composition, stacking fault energy (SFE), and texture [1]. SFE has a significant effect to deformation mechanism, in which metals with medium and high SFE tend to undergo slip mechanism during deformation while on the other hand metals with low SFE experience twinning mechanism other than slip [2]. SFE level also influences the texture formation of metals. During deformation process, metals with high SFE tend to form the copper and S type texture with crystal orientation of {112} <111> and {123} <634>, respectively, which resulted from the slip mechanism during deformation. In contrast, a research showed that metals with low SFE tended to configure the copper type texture at low deformation level that at higher deformation level would rotate and change to brass and Goss type texture at the orientations of {110} <112> and {110} <100>, respectively [2].
Presently existing theories regarding texture transition mechanism and critical deformation level of when the transition occurs in brass alloys are still left in discussion. Hu et al. [3] stated that copper type texture was dominant to form at high temperature process, whereas brass type texture occurred at low temperature. Leffer [4] observed that copper type texture happened at low strain level, further, brass type texture occured at high strain level. Furthermore, Wasserman [5] found that copper type texture promoted by slip mechanism, on the other hand, brass type texture supported by twinning. Carstensen et al. [6] found that copper type texture start to form at the low deformation level, while a research by Hirsch et al. [7] showed that the texture transition begin at 50 % deformation level in which SFE acted predominantly. Verma et al. [8] then stated that texture transition started to form at the deformation level of 15 %. Another factor which appears to influence the deformation mechanism and texture transition is the grain size. Decreasing grain size was found to promote the increase in twinning density [9]. Deformation characteristics are also affected by the presence of second phases which act as pinning agents during dislocation movement which then result in the formation of heterogeneous deformation – shear band – and local deformation zone around the particles [10].
Brass is a type of alloy with low SFE, ranging from 14 - 35 $mJ/m^2$ depending to its Zn content [11]. Higher Zn content results in lower SFE value. Brass alloys have a decent combination of strength and ductility, thus are widely used for applications requiring high mechanical properties such as munition cartridge cases [12]. Addition of alloying element on brass is aimed to improve their physical and mechanical properties that highly depend both on the base alloy composition and the type and amount of alloying element. Previous research showed that the addition of 0.3-2.0 wt. % Si on Cu-10Zn and 0.1 wt. % Co on Cu-10Zn-1.5Si alloy refined the grain size, increased the strength and hardness, while reduced the elongation of brass alloy [1]. Similar grain refinement was found by the addition of 0.1 wt. % Bi to Cu-30Zn, however no significant effect was found on the mechanical properties [13-15]. This research was intended to comprehend the effect of Bi on the microstructures, deformation characteristics, and texture development of Cu-29Zn alloy.

## 2. Experimental Methods

The Cu-29Zn-xBi (wt.%) alloys were produced by gravity casting at pouring temperature of 1150 °C, using pure Cu and Zn ingots, as well as Bi ingot as feeding materials. Bi addition was varied to 0.2, 0.4 and 0.8 wt. %. The metal was melted in a crucible and cast into a 600 °C preheated metal mold with the dimension of 110x110x6 $mm^3$. After casting, the alloys were subjected to homogenization treatment at 800 °C for 2 h in a muffle furnace cooled in air. Composition of produced alloys is shown in Table 1. Speciments were then cold rolled with the level of deformation of 20, 40, and 70 % in multiple passes. Microstructure characterizations were carried out by means of an optical and scanning electron microscopes. For this purpose, samples were prepared using conventional metallographic techniques followed by etching with 10% $FeCl_3$ solution for 5-10 s. Vickers microhardness measurement was performed at regular interval with an applied load of 300 g and a dwell time of 15 s, in accordance with ASTM E387. Slip and twinning densities were defined as the length of slip and twin boundary lines in unit area and were measured at an optical microscope image with the size of 880 x 680 $\mu m^2$. Texture measurement was carried out on as-rolled speciments using Zeiss Ultra Plus SEM-EBSD (Electron Backscatter Diffraction) with step size and scan area of 1 μm and 1000 x 500 $\mu m^2$, respectively. SEM-EBSD data were then analized using Tango Maps and Mambo Pole Figures sofware to obtain pole figure images.

**Table 1**. Nominal composition of the as-homogenized alloys (wt. %).

| Samples | Zn | Bi | Pb | Mn | Fe | Si | Cr | Al | Co | Cu |
|---|---|---|---|---|---|---|---|---|---|---|
| Cu-29Zn | **29.2** | **0.269** | 0.005 | 0.006 | 0.005 | 0.005 | 0.002 | 0.002 | 0.032 | **70.3** |
| Cu-29Zn-0.5Bi | **29.3** | **0.409** | 0.005 | 0.002 | 0.005 | 0.005 | 0.001 | 0.002 | 0.021 | **70.2** |
| Cu-29Zn-1.0Bi | **29.3** | **0.797** | 0.005 | 0.002 | 0.007 | 0.005 | 0.001 | 0.002 | 0.021 | **69.8** |

## 3. Results and Discussion

### *3.1. Microstructures and Hardness*

Figure 1 (a-c) show the as-homogenized Cu-29Zn alloys microstructures with Bi contents of 0.2, 0.4 and 0.8 wt. % obtained using optical microscope while result from SEM observation are referred in Figure 1 (d-f). From Figure 1, it can be seen that gas porosities are present in the samples on account of hydrogen mix-up during casting process. The Cu-29Zn alloys with Bi addition show both single phase (α) and Bi dispersoid. The dispersoid are richer in Bi compare to that found in the matrix. This phase is found segregated in the grain and along the grain boundary in globular forms. Bi is completely insoluble in brass thus tends to segregate to the solid-liquid interface during solidification.

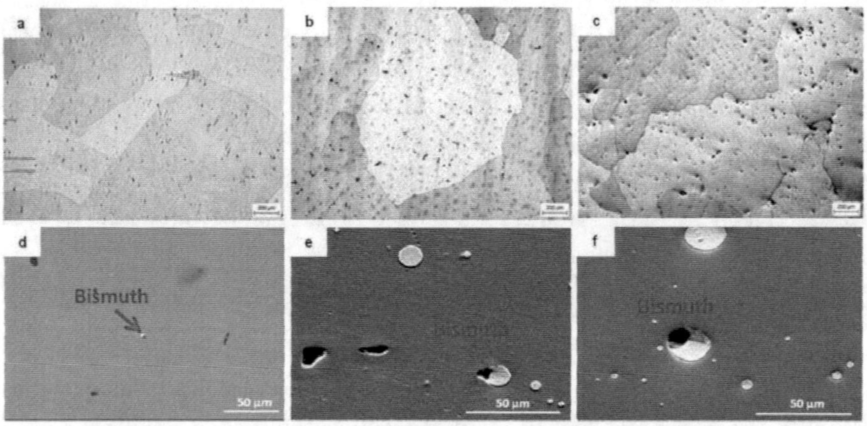


**Figure 1.** Micrographs of as-homogenized Cu-29Zn alloys with (a,d) 0.2 [16] (b,e) 0.4 (c,f) 0.8 wt. % Bi.

Figure 2 (a-i) illustrates the microstructure of Cu-29Zn, Cu-29Zn-0.4Bi, and Cu-29Zn-0.8Bi alloys after cold rolling with deformation level of 20, 40, and 70 %. Results of slip and twin density measurements are provided in Figure 3. At 20 % deformation, slip acted dominantly and the presence of Bi was found to increase the slip density. The slip densities of Cu-29Zn alloys with Bi addition of 0.2, 0.4, and 0.8 wt. % are $33 \times 10^{-3}$, $61 \times 10^{-3}$ dan $91 \times 10^{-3}$ $\mu m^{-1}$, respectively (Figure 3). Bi atoms acted as pinning agents during solidification which inhibited the movement of atoms and resulted in smaller grain size. During deformation, Bi atoms interfered with the slip so that increased the slip density. At 40 % deformation, twinning replaced slip as the predominant mechanism and its density slightly increased with the presence of Bi. The Bi dispersoid retarded the homogeneous deformation and promoted inhomogeneous deformation [16]. Further deformation at 70 % produced shear bands and flattened the Bi dispersoid. Greater Bi content was then found to induce the formation of shear band. Figure 4 shows the effect of Bi content and deformation level on the hardness of Cu-29Zn alloys. The figure shows that the increase of Bi content have no significant effect on hardness. Meanwhile, the alloy hardness increased with the increase in level of deformation caused by strain hardening.

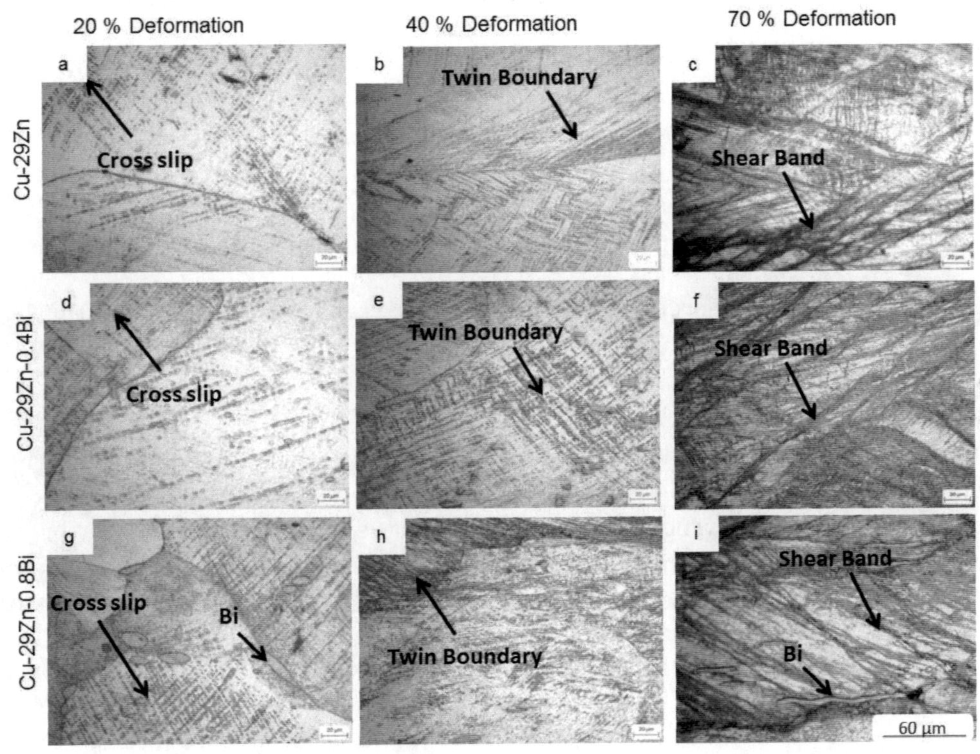


**Figure 2**. Micrographs of Cu-29Zn alloy with the Bi content of (a-c) 0.2 [16] (d-f) 0.4 (g-i) 0.8 wt.% after cold rolling for 20, 40, and 70% thickness reduction.

IOP Conf. Series: Materials Science and Engineering **378** (2018) 012015 doi:10.1088/1757-899X/378/1/012015

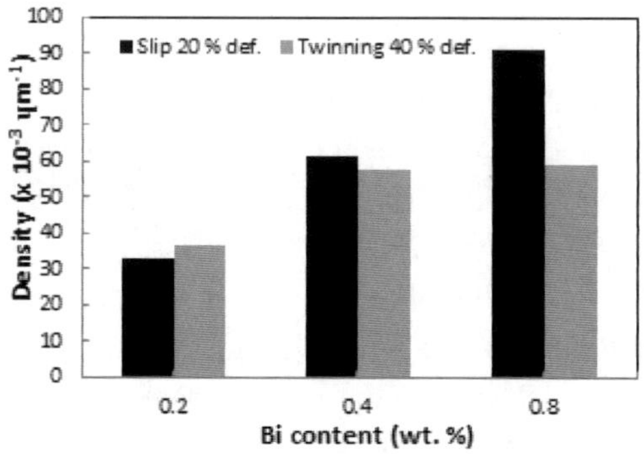


**Figure 3**. The effects of Bi content on the slip and twinning density of Cu-29Zn alloys at the deformation levels of 20 and 40 %.

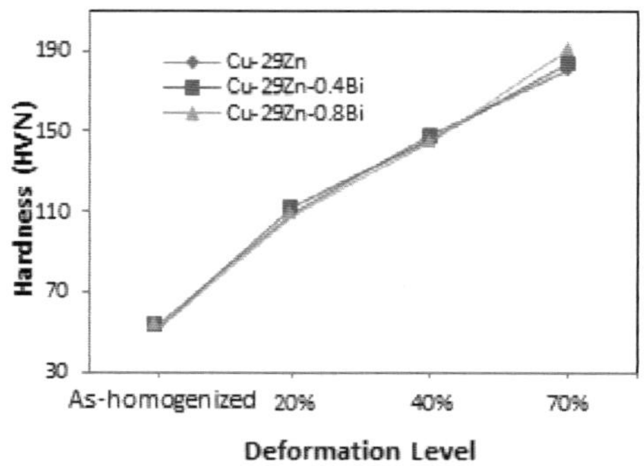


**Figure 4.** The effect of Bi content and deformation level on hardness of Cu-29Zn alloys.

*3.2. Texture*

Figure 5 (a-i) illustrates the EBSD images of Cu-29Zn, Cu-29-0.4Bi, and Cu-29Zn-0.8Bi alloys after cold rolling with deformation level of 0, 40, and 70 %. These images were obtained from analysis using Tango Maps software in which different colour indicated different crystal orientation. Figure 6 (a-i) shows the pole figure images that obtained from analysis using Mambo Pole Figures software. From Figure 6 (a,d,g), it is clear that the addition of Bi on Cu-29Zn alloy has no significant effect on homogenization texture in which all samples showed the random texture.

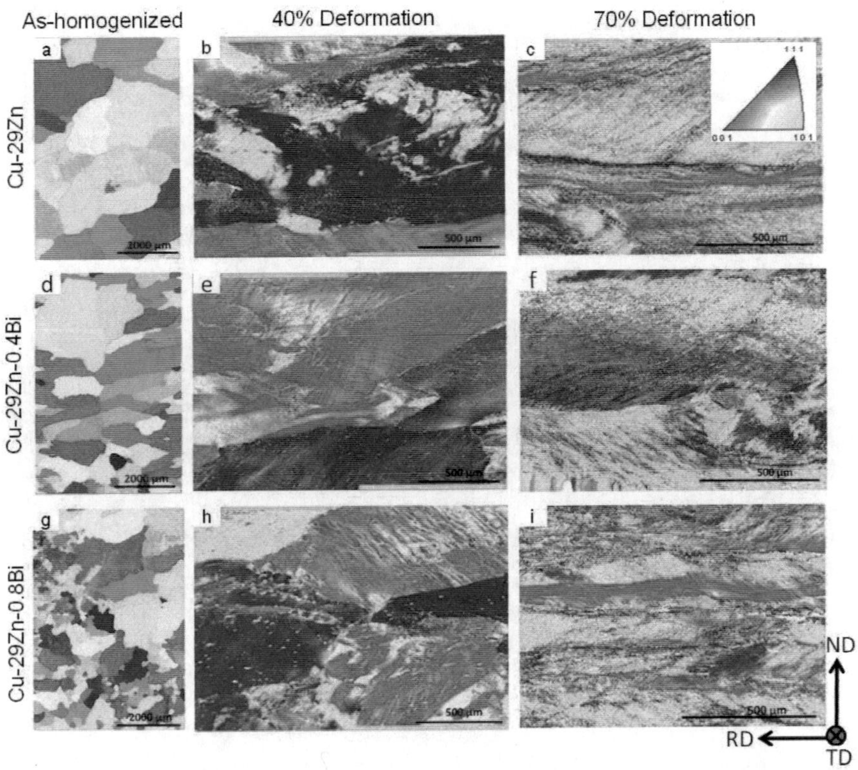


**Figure 5**. EBSD images of Cu-29Zn alloys with (a) 0.2 (d) 0.4 (g) 0.8 wt. % Bi after homogenization process and cold rolling for 40, and 70 % thickness reduction.

At 40% deformation, pole figure images (Figure 6 (b,e,h)) show different pattern when compared with the homogenized samples. They showed the trend of copper type texture with orientation of {112} <111> with intensity of 7.92. This phenomena is similar with previous observation found by Duggan et al. [17] and Hutchinson and Ray [18] that Cu-30Zn alloy with 40-60% deformation showed strong copper type texture in orientation of {112} <111>. This texture formed due to mechanical twinning during deformation process. At 70 % deformation, Figure 6 (c,f,i) illustrated the combination of brass and Goss type texture at orientation of {110} <112> and {110} <001>, respectively. At this deformation level, the copper type texture rotated to brass and Goss type texture due to the shear band [18]. Addition of Bi to Cu-29Zn alloys increased the shear band density due to the Bi dispersoid which acted as the pinning agent during homogeneous deformation and increased the inhomogeneous deformation as shear band. The shear band mechanism promoted the formation of brass type texture [18].

IOP Conf. Series: Materials Science and Engineering **378** (2018) 012015 doi:10.1088/1757-899X/378/1/012015

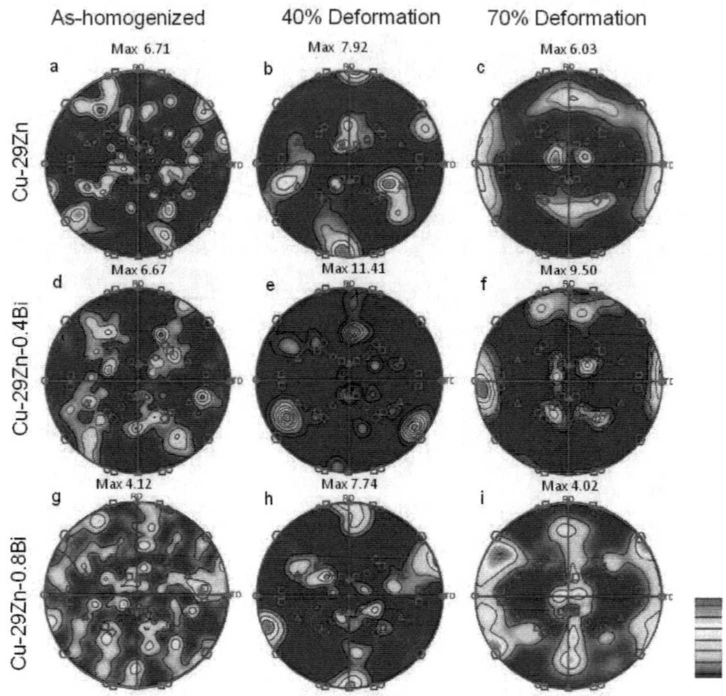


Copper {112}<111> △, Brass {110}<112> ○, Cube {100}<001> ◇, Goss {110}<001> ○, S {123}<634> □

**Figure 6**. Pole figure images of Cu-29Zn alloys with (a) 0.2 (d) 0.4 (g) 0.8 wt. % Bi after homogenization process and cold rolling for 40, and 70 % thickness reduction.

## 4. Conclusion

The results of the observation can be concluded as follows:

a) Addition of Bi to Cu-29Zn alloy promoted the formation of Bi dispersoid which segregated in the grain and along the grain boundary in globular forms. Bi is completely insoluble in brass and thus tends to segregate to the solid-liquid interface during solidification and form isolated particles inside the grain and along the grain boundaries.

b) Slip mechanism was dominant at 20% deformation level and the presence of Bi increased the slip density. During deformation, Bi atoms interfered with the slip so that increased the slip density.

c) Twinning was found as the corresponding plastic deformation mechanism at deformation level of 40% and its density slightly increased with the presence of Bi. The Bi dispersoid acted as a retardant to the homogeneous deformation furthermore promoted the occurence of inhomogeneous deformation. At 70 % deformation, the appearance of shear band was dominant with flattened Bi dispersoid and filled intergranular space.

d) The addition of Bi on Cu-29Zn alloys has no significant effect on the texture of as-homogenized samples texture of which all showed random patterns. At 40% deformation, pole figure images showed the trend of copper type texture with orientation of {112} <111> with intensity of 7.92. At 70 % deformation, all samples illustrated the combination of brass and Goss type texture at orientation of {110} <112> and {110} <001>, respectively. Addition of Bi on Cu-29Zn alloys increased shear band density due to the occurrence Bi dispersoid acting as the pinning agent during homogenous deformation that increased the inhomogeneous deformation as shear band. The shear band mechanism promoted the formation of brass type texture.

## 5. Acknowledgement

This research was funded by Advanced Research Grant (Hibah Penelitian Unggulan Perguruan Tinggi, PUPT) 2017 from Universitas Indonesia. The authors thanks Prof. Simon P. Ringer for the provision of facilities as well as scientific and technical assistance at the Australian Microscopy & Microanalysis Research Facility of the Australian Centre for Microscopy & Microanalysis, The University of Sydney, Australia.

## 6. References

[1] Oishi K., Sasaki I., Otani J., Effect of Silicon Addition on Grain Refinement of Copper Alloys, Mat. Letter, Vol. 57, 2003, pp. 2280 - 2286.

[2] Yan H., Zhao X., Jia N., Zheng Y., He T., Influence of Shear Banding on the Formation of Brass-type Textures in Polycrystalline FCC Metals with Low Stacking Fault Energy, J. Mater. Sci. Tech. Vol. 30, 2014, pp. 408-416.

[3] Hu H., Goodman S.R., Texture Transition in Copper, Trans. Am Inst. Min. Metall. Pet. Eng., Vol. 227, 1963, pp. 627-639.

[4] Leffer T., Deformation Rate Dependence of Rolling Texture in Brass Containing 5% Zn, Scripta Metall., Vol. 2, 1968, pp. 447-452.

[5] Wasserman G., The Effect of Mechanical Twinning on the Formation of Rolling Textures in Faced-Centered-Cubic Metals, J. Metall, Vol. 64, 1963, pp. 61-65.

[6] Carstensen J.V., Ray R.K., Leffers T., Texture Development in Ni-Co Alloys Rolled to Moderate Reductions, Mat. Sci. Forum, Vol. 408, 2002, pp. 607-612.

[7] Hirsch J., Lucke K., Hatherly M., Mechanism of Deformation and Development of Rolling Texture in Polycrystalline FCC Metals - III. The Influence of Slip Inhomogeneities and Twinning, Acta Met., Vol. 36, 1988, pp. 2905-2927.

[8] Verma A.K., Shingweker A., Nihichlani M., Singh V., Mukhopadhyay P., Deformation Characterization of Cartridge Brass, Indian J. of Eng. & Mat. Sci., Vol. 20, 2013, pp. 283-288.

[9] Li Y., Zhao Y.H., Liu W., Xu C., Horita Z., Liao X.Z., Zhu Y.T., Langdon T.G., Lavernia E.J., Influence of Grain Size on the Density of Deformation Twins in Cu–30%Zn Alloy, Mat. Sci. and Eng. A, Vol. 527, 2010, pp. 3942-3948.

[10] Suwas S. and Ray R.K., Crystallographic Texture of Material, 2014, Springer, London.

[11] Wu X., Wen C., Gong Y., Ren S., Tao J., Chen L., Long Y., Zhu X., Effect of Stacking Fault Energy and Strain Rate on the Mechanical Properties of Cu and Cu Alloys, J. Alloys and Compound, Vol. 573, 2013, pp. 1-5.

[12] Feng C., Clark S., Malfunction and Failure Analysis Investigations of C26000 (Cu-30% Zn) Brass Cartridge Cases, Mat. Characterization, Vol. 32, 1994, pp. 15-23.

[13] Emelina N.B., Alabin A.N., Belov N.A., Influence of Bismuth and Lead on the Formation of the Structure of Experimental Alloys with a Cu–30% Zn Based Composition during

Crystallization, Deformation, and Thermal Treatment, Russian J. Non Ferrous Metals, Vol. 51, 2010, pp. 476–482.

[14] Basori I., Gadhu R., Sofyan B.T., Effects of Deformation and Annealing Temperature on the Microstructures and Hardness of Cu-29Zn-0.6Bi Brass, Key Eng. Mat., Vol. 748, 2017, pp. 218-223.

[15] Basori I., Jendra D., Sofyan B.T, Effect of Bi Addition on the Mechanical Properties of Cu-29Zn Alloy, Proceeding of the National Seminar on Mechanical and Vocational Engineering (Prosiding Seminar Nasional Mesin dan Teknologi Kejuruan), Jakarta, 2016, pp. 274-277 ( in Bahasa Indonesia).

[16] Basori I., Hafsah I., Sofyan B.T., Effect of Manganese on the Microstructure, Mechanical properties and deformation characteristic of Cu-29Zn alloy, Mat. Sci. Forum, 2017, in-press.

[17] Duggan B.J., Hatherly M., Hutchinson W.B, Wakefield P.T., Deformation Structures and Textures in Cold-Rolled 70:30 Brass, Met. Sci., Vol. 12, 1978, pp.343-351.

[18] Hutchinson W.B, Ray R.K., Influenced of Phosporus Addition on Annealing Behaviour of Cold-Work Copper, Met. Sci., Vol. 13, 1979, pp. 125-130.

IOP Conf. Series: Materials Science and Engineering **378** (2018) 012016 doi:10.1088/1757-899X/378/1/012016

# Effects of material characteristics and clearance size on dynamics of a slider-crank mechanism with two sliders and revolute clearance joints

**Ngoc-Thai Huynh[1], Shyh-Chour Huang[1*], Thanh-Phong Dao[2,3]**

[1]Department of Mechanical Engineering, National Kaohsiung University of Applied Sciences, 415 Chien-Kung Road, Kaohsiung 80778, Taiwan R.O.C.
[2]Division of Computational Mechatronics, Institute for Computational Science, Ton Duc Thang University, Ho Chi Minh City, Vietnam
[3]Faculty of Electrical & Electronics and Engineering, Ton Duc Thang University, Ho Chi Minh City, Vietnam

Email: shuang@cc.kuas.edu.tw

**Abstract.** This article presents influences of clearance size and material parameters on the dynamic response of a slider-crank mechanism with imperfect revolute joints in dry contacting condition. The mechanism was created by Solidworks and a finite element method in ANSYS software was used to analyze the effects which presented and discussed. The simulation results revealed that the acceleration of two sliders was obviously shaking with high peaks when the clearance size equal to 0.3 mm which differ from the ideal joint. Moreover, the smaller Young's modulus material of journals indicated the significant effects on acceleration of mechanical systems. The reasons are due to a suddenly increase of contact force when the journal impacted into the bearing in imperfect revolute joints. The simulation results for the clearance size equal to 0.1 mm, 0.2 mm and 0.4 mm were close to an ideal joint. It demonstrated that clearance size and material characteristic played an important role in dynamics analysis of mechanical systems.

**Keywords**: Material characteristic, slider-crank mechanism, two sliders, revolute clearance joint, dynamics.

## 1. Introduction

There are some significant parameters that cause low mechanical accuracy for a slider-crank mechanism with two sliders. First, a clearance in a revolute joint is inevitable due to the design, assembly, and fabrication tolerances. The clearance causes undesired vibration and abrasion of the joint which increase the clearance size. Second, the material characteristics which caused friction and lead to an abrasion and a deformation of the joint. Moreover, the clearance size causes a contact-impact force in the revolute joint, which leads to a low accuracy in the mechanical system. The impact force also causes an abrasion joint and increases the clearance size. The different driving speed influences on the dynamic behaviors. The effect of the clearance size was investigated and presented by researchers which based on some contact force models as presented in [1] , the simulation result compared with the experimental data with the different clearance size and driving speed. Besides, the flexible components and two clearance joints also were considered and presented in [2], which wear prediction and optimization of maximum allowable wear occurs at the same number of rotation cycles for two clearance joints simultaneously. Moreover, the effect of multiple clearance joints on dynamic of planar multibody mechanical system were analyzed and presented in [3], in which it was shown three motion phases of journal: free motion, contact motion and impact motion, comparison between ideal joint and one clearance joint, two clearance, three clearance joint. In order to restrict clearance joint effects, a compliant mechanism or flexible mechanism was investigated in [4], in which Taguchi method was used in this paper. However, the different material characteristic between journal and bearing causes friction force lead to joint wear and then induced damage for mechanical systems as reported in [5], in which influences of restitution coefficient and material characteristic cannot be ignored.

The objective of this paper is to investigate effects of clearance size and material characteristic on the dynamic of slider-crank mechanism with two sliders. Dynamic simulations were analyzed by using the finite element analysis in ANSYS. The results and discussion were presented.

## 2. Modeling

### *2.1. The slider-crank mechanism*

The slider-crank mechanism with two sliders and seven revolute clearance joint was used to analyze, the model is shown Fig. 1. The mechcanism consists of the rigid links namely (1) the 1st connecting rod, (2) the 2nd connecting rod, (3) the 1st ray, (4) the 1st slider, (5) the 2nd ray, (6) crank, (7) motor, (8) Base, (9) the 2nd slider, (10) rotation of motor control, (11) DC.

In order to capture the clearance in revolute, Fig. 2 depicts a scheme of a planar slider-crank mechanism with two sliders. It includes revolute clearance joints $R_1$, $R_2$ and $R_3$ exist between crank, the 1st connecting rod and the 2nd connecting rod, two revolute clearance joints $R_4$ and $R_5$ exist between connecting rod 1st and slider 1, two revolute clearance joints $R_6$ and $R_7$ exist between the 2nd connecting rod and the 2nd slider, one ideal revolute joint exist between base and motor shaft, and two translations joint $T_1$ and $T_2$ exist between the 1st ray and the 1st slider, between the 2nd ray and the 2nd slider. These parameters were listed in Table 1. The crank is fixed to motor shaft. Two sliders are driven from motor shaft's rotation through the crank, the 1st connecting rod and 2nd connecting rod to the 1st slider and the 2nd slider, turn the crank's rotation into linear motion of the 1st slider and the 2nd slider.

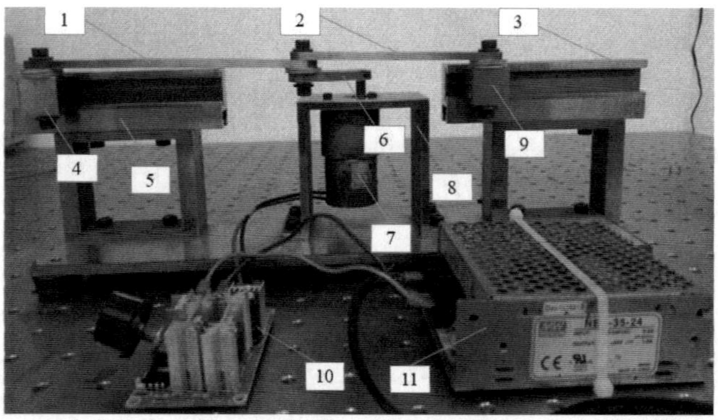


**Figure 1.** Rigid Sider-crank Mechanism ( (1) The 1st connecting rod, (2) The 2nd connecting rod (3) The 1st ray, (4) The 1st slider , (5) The 2nd ray, (6) Crank, (7) Motor, (8) Base, (9) The 2nd slider, (10) Rotation of motor control, (11) DC

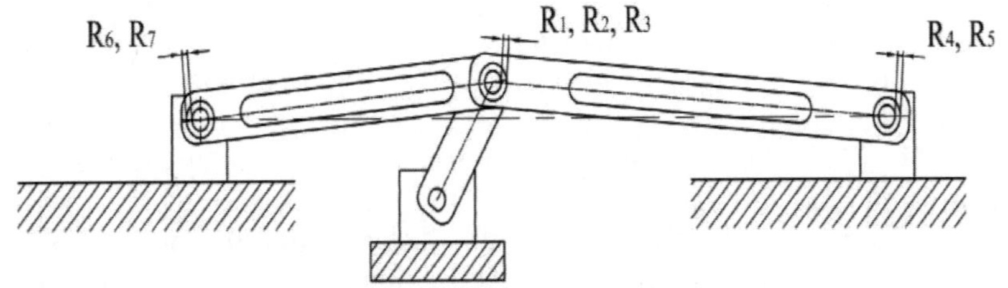


**Figure 2.** Planar Slider-crank mechanism with revolute clearance joint

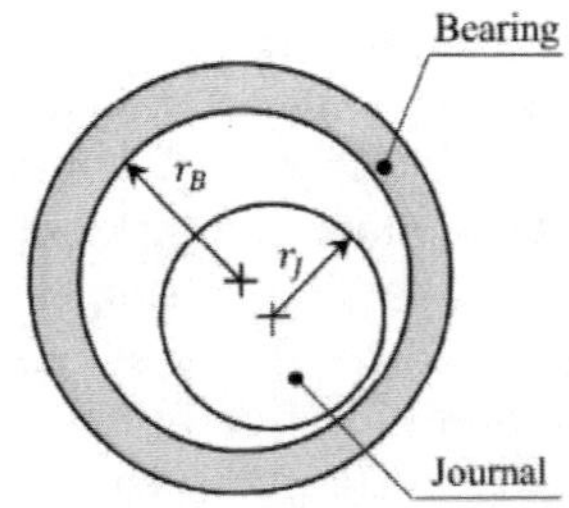


**Figure 3.** Revolute joint with clearance

**Table 1.** Simulations characteristics

| Simulations characteristics | Value | Simulations characteristics | Value |
|---|---|---|---|
| Modulus of elasticity (GPa) | 207 | Length of connecting rod 2nd (mm) | 120 |
| Density (kg/m$^3$) | 7850 | Mass of crank (kg) | 0.03 |
| Poisson's ratio | 0.3 | Mass of connecting rod 1st (kg) | 0.064 |
| Initial length of crank (mm) | 50 | Mass of connecting rod 2nd (kg) | 0.047 |
| Length of connecting rod 1st (mm) | 160 | Mass of slider 1st and 2nd (kg) | 0.092 |

### 2.2. *Modeling of revolute clearance Joint*

In general, it is well known that two bodies are linked by joint which are called journal-bearing. In fact, the existence of clearance in journal-bearing joint is inevitable. The revolute clearance joint was depicted in Fig. 3. The difference between journal and bearing is radial clearance joint and it is defined as follows:

$$c = r_b - r_j \tag{1}$$

where $r_b$, $r_j$ are the radii of bearing and journal, respectively.

Although, a revolute clearance joint does not restrict any degree of freedom of a mechanical system like the ideal joint, it imposes some kinematic restrictions and limiting the journal to move within the bearing. Thus, when the clearance size exists in a revolute joint, it introduces two extra degrees, which are horizontal and vertical displacements.

## 3. Results and discussion

### 3.1. *Effects of different clearance size on dynamics*

The motor shaft is a driving link and rotates with a constant velocity equal to 600 rpm with five different clearance size equal to zero, 0.1 mm, 0.2 mm, 0.3 mm and 0.4 mm, respectively.

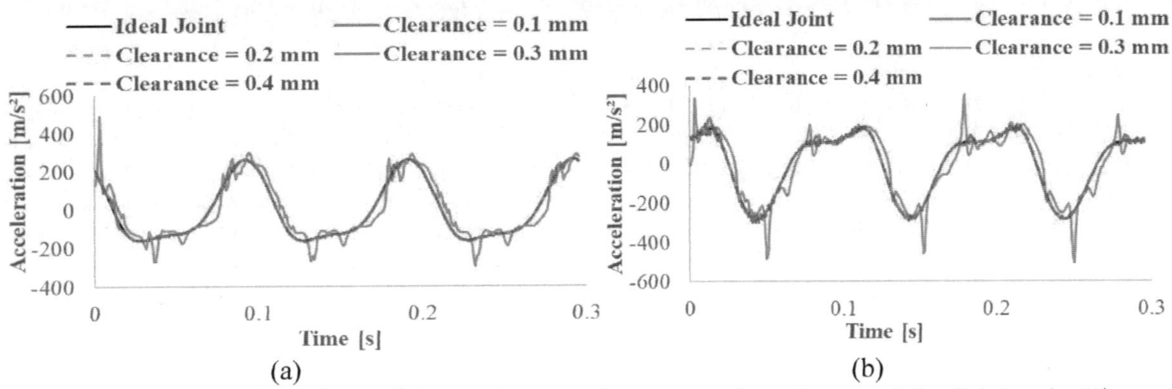


**Figure 4.** Accelerations of two sliders and contact force in revolute clearance joint ((a) For the 1st slider, (b) For the 2nd slider

The slider-crank mechanism with two sliders and seven revolute clearance joints was described in section 2, using finite element analysis in ANSYS. Radius of bearing equal to 4.5 mm, radius of journal is 4.4 mm, 4.3 mm, 4.2 mm and 4.1 mm, respectively. Initialy, the journal and bearing are contact with each other, duration time simulation is 1s, time step is $10^{-4}$ and simulation results preseted in duration time 0.3s, in Figures. (4-5). It is identified that clearance size has significantly effected on accelerations of two sliders which oscillated with hight peaks when the clearance size equals to 0.3 mm and has silghtly effected when clearance size equals to 0.1 mm, 0.2 mm and 0.4 mm. This is because the curves of acceleration are close to the case ideal joint. At the time, accelerations have high peaks due to the journal impact into the bearing that causes the contact force strongly incereased and pointed out in Figure. 5(a)-(b)-(c)-(d). The contact force curves were different from the ideal joint case. The same phenomena can be found in [1-3]. From the simulation results were found that the clearance size cannot be ignored when analyzed dynamic of multi-body mechanical sysytems.

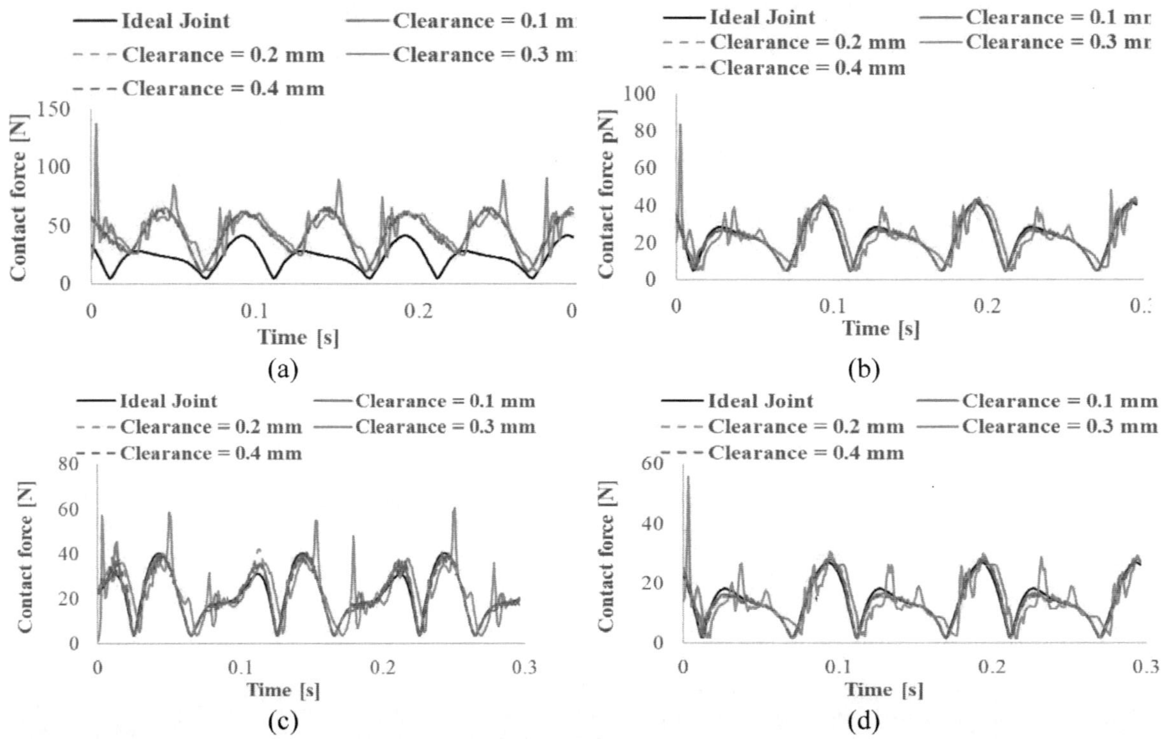


**Figure. 5.** The contact force in revolute clearance joint ((a) Contact force of $R_1$, (b) Contact force of $R_2$, (c) Contact force of $R_3$, (d) Contact force of $R_4$)

### *3.2. Effects of different material characteristics on dynamics*

The motor shaft is a driving link and rotates with a constant velocity equal to 600 rpm. Various materials are used being manufactured for the journal. For example, Young's modulus of structure steel, iron, copper and aluminum Al-6061-T6 is 207 GPa, 173 GPa, 106 GPa and 69 GPa, respectively. Poisson's ratio of the corresponding materials is 0.3, 0.3, 0.324 and 0.33, respectively. The radius of bearing and journal equals to 4.5 mm and 4.2 mm, respectively. The clearance size equals to 0.3 mm and at the initial time, journal and bearing contacted with each other. The material of bearing is steel, the material of journal in ideal joint is SS-304 with Young's modulus 193 GPa and Poisson's ratio is 0.3. The simulation results pointed out in Figures (6-7).

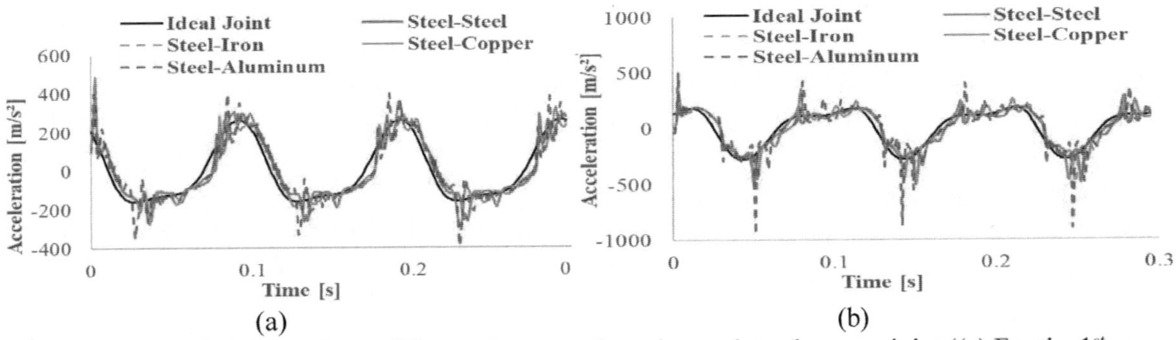


**Figure 6.** Accelerations of two sliders and contact force in revolute clearance joint ((a) For the 1st slider, (b) For the 2nd slider

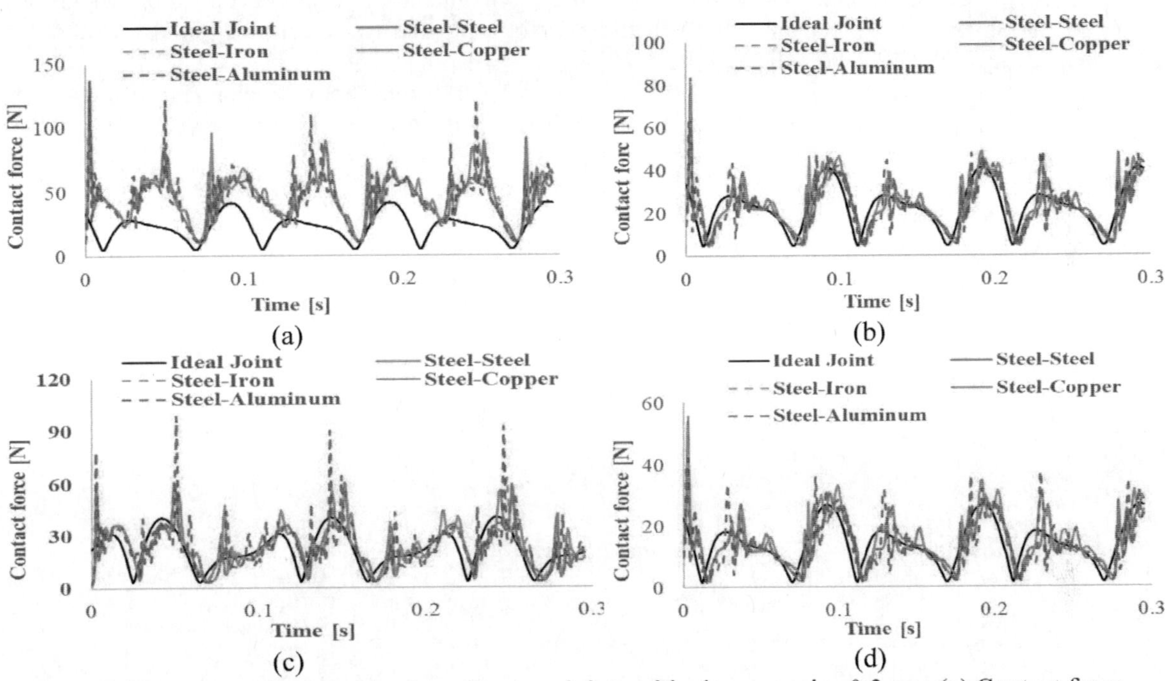


**Figure 7.** The contact force in revolute clearance joints with clearance size 0.3 mm (a) Contact force of $R_1$, (b) Contact force of $R_2$, (c) Contact force of $R_3$, (d) Contact force of $R_4$)

In Fig. 6(a)-(b), it is clear that the accelerations curves of two silders with steel-aluminum, steel-copper has obvious shaking with high peaks compared with steel-steel and steel-iron. Meanwhile, the result of steel-aluminum has slighter delay than that of steel-copper, and steel-iron's results have slighter delay than steel-steel's. The reason for this phenomenon is that according to the material combination of steel-steel, steel-iron, steel-copper and steel-aluminum, the contact stiffness is gradually decreased. The same phenomena can be found in the contact force curves as depicted in Figure. 7 (a)-(b)-(c)-(d). The results was in line with present in reference [5].

## 4. Conclusions

The influences of non-ideal revolute joints on the dynamic behavior of a slider-crank mechanism with two sliders were analyzed using the Finite element method. the accelerations of the two sliders were obviously shaking with high peaks due to the sudden impact of journal into bearing that causes contact force in the revolute clearance joint suddenly increase when clearance size equal to 0.3 mm. When the clearance size equals to 0.1 mm, 0.2 mm and 0.4 mm has slightly affected on the accelerations of two sliders and it is close to ideal joint. But the contact force was different from ideal joint. Further, different material characteristics have important effected on dynamic response of slider-crank mechanism with two sliders when the clearance size equals to 0.3 mm. Material has Young's modulus smaller cause high peaks for the accelerations and contact force. So that, it will cause the joints wear quickly and induce damage for mechanical systems.

## 5. Acknowledgments

The authors acknowledge and thank the Ministry of Science and Technology of the Republic of China for their partial financial support of this study under Contract Number MOST 105-2221-E-151-016.

## 6. References

[1] J. Ma, L. Qian, G. Chen, and M. Li, "Dynamic analysis of mechanical systems with planar revolute joints with clearance," *Mechanism and Machine Theory,* vol. 94, pp. 148-164, 2015.

[2] Y. Li, G. Chen, D. Sun, Y. Gao, and K. Wang, "Dynamic analysis and optimization design of a planar slider–crank mechanism with flexible components and two clearance joints," *Mechanism and Machine Theory,* vol. 99, pp. 37-57, 2016.

[3] Z. F. Bai and Y. Sun, "A study on dynamics of planar multibody mechanical systems with multiple revolute clearance joints," *European Journal of Mechanics - A/Solids,* vol. 60, pp. 95-111, 2016.

[4] S.-C. H. Thanh-Phong Dao, "Design and Analysis of Flexible Slider Crank Mechanism," vol. 8, pp. 836-843, 2014.

[5] X. Wang, G. Liu, S. Ma, and R. Tong, "Effects of restitution coefficient and material characteristics on dynamic response of planar multi-body systems with revolute clearance joint," *Journal of Mechanical Science and Technology,* vol. 31, pp. 587-597, 2017.

IOP Conf. Series: Materials Science and Engineering **378** (2018) 012017 doi:10.1088/1757-899X/378/1/012017

# Dynamic Characteristics Analysis of Spiral Groove Liquid Lubricated Seal Considering Centrifugal Inertia

**W J Yang, M M Hao, Z T Li and H C Cao**

College of Chemical Engineering, China University of Petroleum (East China), Qingdao, 266580, China

E-mail: yangwen516@126.com

**Abstract.** A three degree of freedom model of spiral groove liquid lubricated seal was established for studying the effects of centrifugal inertia on dynamic characteristics. The dynamic stiffness and damping coefficients of liquid lubricated seal with centrifugal inertia and those without centrifugal inertia were obtained by means of finite element method, and then comparative analysis was carried out. Results indicate that dynamic stiffness and damping coefficients are increased linearly with the increase of rotating speed. Dynamic stiffness and damping coefficients decrease with the increasing film thickness. The effects of centrifugal inertia on axial stiffness coefficient and angular coupling stiffness coefficient can be neglected under different rotating speed and different film thickness. When the values of rotating speed and film thickness are large, the effect of centrifugal inertia on angle stiffness coefficient cannot be ignored. Centrifugal inertia has no effects on dynamic damping coefficients with the variation of rotating speed and film thickness.

## 1. Introduction

The spiral groove liquid lubricated mechanical seal has been widely applied in the field of rotary machinery, such as pumps, compressors, mixers and so on. Its configuration is shown in figure 1. Compared with a conventional mechanical seal, the surface of rotating ring contains a pattern of shallow spiral grooves, which generate the hydrodynamic pressure to separate the sealing rings [1]. With the increasing rotating speed, a great deal of work has been done on centrifugal inertia. However, most of the work has dealt with bearings [2-3]. The Reynolds equation incorporating centrifugal forces was derived by Pinkus and Lund [4], and then an analysis was conducted and solutions were provided for the effect of centrifugal forces on the hydrodynamics of high-speed thrust bearings and seals. Gupta and Sharma presented the analysis of centrifugal inertia effects in fluid-film-lubricated misaligned radial face seals [5]. Nowadays, most researchers have neglected the centrifugal inertia but Zhao did some work on spiral-groove rotary seal ring with considering the effect of centrifugation [6-7]. The ability to eliminate or minimize direct face contact during runtime is the key to the success of a liquid lubricated face seal. If the stator cannot track the rotor properly, seal face contact may occur, which would lead to face wear or seal failure [8]. Thus, it is important to study the effect of centrifugal inertia on dynamic characteristics of liquid lubricated seals.

In this paper, a three degree of freedom model of spiral groove liquid lubricated seal is established. The stiffness and damping coefficients of liquid lubricated seal with centrifugal inertia and those without centrifugal inertia are obtained by the perturbation method and the finite element method. Then comparative analysis is carried out. The results can provide theoretical guidance for the design and application of liquid lubricated seal under different operating conditions.

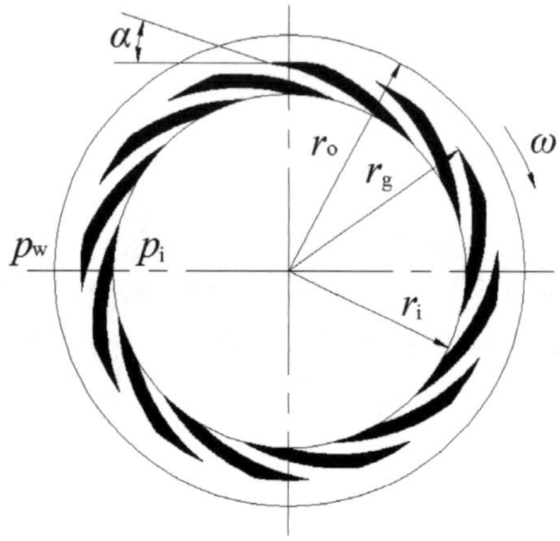


**Figure 1.** Geometry schematic diagram of rotating ring.

The following analysis is based on the parameters listed in table 1, which includes seal geometry and operating conditions.

**Table 1.** Parameters for the analysis of a spiral groove liquid lubricated seal

| Parameters | Working conditions |
|---|---|
| inner radius $r_i$(m) | 0.04425 |
| outer radius $r_o$(m) | 0.05325 |
| spiral groove radius $r_g$(m) | 0.05055 |
| spiral groove depth $h_g$(m) | $12.5\times10^{-6}$ |
| spiral groove angle $\alpha$(°) | 18 |
| spiral groove number $N_g$ | 12 |
| land-to-groove ratio | 1 |
| lubricant velocity $\mu$(Pa·s) | 0.001003 |
| lubricant temperature (℃) | 25 |
| lubricant density $\rho$(kg·m$^{-3}$) | 1000 |
| inner pressure $p_i$(MPa) | 0.6 |
| outer pressure $p_w$(MPa) | 0.4 |

## 2. Governing equations

To establish the mathematical model, some basic assumptions are made as following.

- The fluid between the sealing faces can be characterized as Newtonian fluid, and it is isothermal and isoviscous. Also, the fluid film is laminar.
- The seal face is smooth and no face contact happens in the operation.

The Reynolds equation considering centrifugal inertia can be expressed in the dimension form as follows [9]:

$$\frac{\partial}{\partial r}\left[\frac{r\rho h^3}{\mu}\left(\frac{\partial p}{\partial r}\right)\right]+\frac{1}{r}\frac{\partial}{\partial\theta}\left[\frac{\rho h^3}{\mu}\left(\frac{\partial p}{\partial\theta}\right)\right]=6r\omega\frac{\partial(\rho h)}{\partial\theta}+12r\frac{\partial(\rho h)}{\partial t}+\frac{3\omega^2}{10\mu}\frac{\partial\left(\rho^2r^2h^3\right)}{\partial r} \tag{1}$$

When the seal is perturbed by a small oscillatory motion ($\Delta z$, $\Delta\gamma_x$, $\Delta\gamma_y$) with frequency $f$, according to the kinematics shown in figure 2, the seal displacement can be expressed as the following:

$$h=h_0+\Delta z+r\sin\theta\Delta\gamma_x-r\cos\theta\Delta\gamma_y \tag{2}$$

where $\Delta z=|\Delta z|e^{ift}$, $\Delta\gamma_x=|\Delta\gamma_x|e^{ift}$, $\Delta\gamma_y=|\Delta\gamma_y|e^{ift}$. $h_0$ is the steady-state equilibrium film thickness.

IOP Conf. Series: Materials Science and Engineering **378** (2018) 012017 doi:10.1088/1757-899X/378/1/012017

These small motions cause a small pressure perturbation about the equilibrium pressure $p_0$, and then the total pressure can be expressed as

$$p = p_0 + p_z\Delta z + p_x\Delta\gamma_x + p_y\Delta\gamma_y \tag{3}$$

Because the motions are defined in complex form, the corresponding perturbation pressures are also complexes, which can be written as

$$p_z = p_{zr} + \mathrm{i}p_{zi}\ ,\quad p_x = p_{xr} + \mathrm{i}p_{xi}\ ,\quad p_y = p_{yr} + \mathrm{i}p_{yi} \tag{4}$$

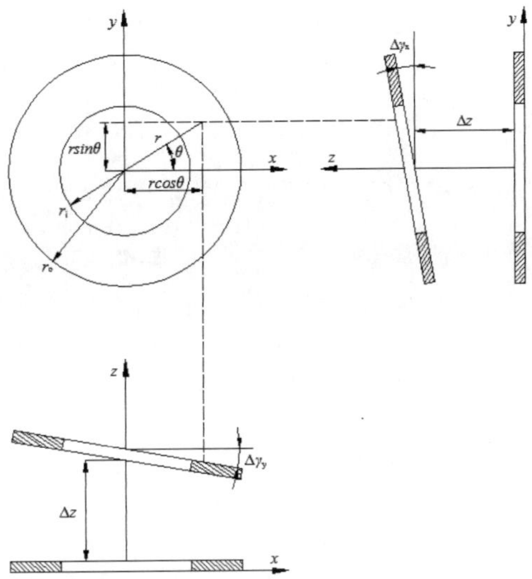


**Figure 2.** Seal perturbation model andcoordinate system.

Substituting relationships from equation (2) to equation (4) into equation (1) and neglecting higher order terms yields the following seven dimensionless equations:

$$\frac{\partial}{R\partial R}\left(RH^3\frac{\partial\overline{P}}{\partial R}\right)+\frac{1}{R^2}\frac{\partial}{\partial\theta}\left(H^3\frac{\partial\overline{P}}{\partial\theta}\right)=K_1\frac{\partial H}{\partial\theta}+K_2\frac{\partial\left(R^2H^3\right)}{R\partial R} \tag{5}$$

$$\frac{\partial}{R\partial R}\left(RH^3\frac{\partial\overline{P}_{zr}}{\partial R}\right)+\frac{\partial}{R\partial R}\left(3RH^2\frac{\partial\overline{P}}{\partial R}\right)+\frac{1}{R^2}\frac{\partial}{\partial\theta}\left(H^3\frac{\partial\overline{P}_{zr}}{\partial\theta}\right)+\frac{1}{R^2}\frac{\partial}{\partial\theta}\left(3H^2\frac{\partial\overline{P}}{\partial\theta}\right)=K_2\frac{\partial\left(3R^2H^2\right)}{R\partial R} \tag{6a}$$

$$\frac{\partial}{R\partial R}\left(RH^3\frac{\partial\overline{P}_{zi}}{\partial R}\right)+\frac{1}{R^2}\frac{\partial}{\partial\theta}\left(H^3\frac{\partial\overline{P}_{zi}}{\partial\theta}\right)=2K_1\omega_1 \tag{6b}$$

$$\begin{aligned}&\frac{\partial}{R\partial R}\left(RH^3\frac{\partial\overline{P}_{xr}}{\partial R}\right)+\frac{\partial}{R\partial R}\left(3H^2R^2\sin\theta\frac{\partial\overline{P}}{\partial R}\right)+\frac{1}{R^2}\frac{\partial}{\partial\theta}\left(H^3\frac{\partial\overline{P}_{xr}}{\partial\theta}\right)+\frac{1}{R^2}\frac{\partial}{\partial\theta}\left(3H^2R\sin\theta\frac{\partial\overline{P}}{\partial\theta}\right)\\&=K_1\frac{\partial\left(R\sin\theta\right)}{\partial\theta}+K_2\frac{\partial}{R\partial R}\left(3H^2R^3\sin\theta\right)\end{aligned} \tag{7a}$$

$$\frac{\partial}{R\partial R}\left(RH^3\frac{\partial\overline{P}_{xi}}{\partial R}\right)+\frac{1}{R^2}\frac{\partial}{\partial\theta}\left(H^3\frac{\partial\overline{P}_{xi}}{\partial\theta}\right)=2K_1\omega_1R\sin\theta \tag{7b}$$

$$\frac{\partial}{R\partial R}\left(RH^3\frac{\partial\bar{P}_{yr}}{\partial R}\right)+\frac{\partial}{R\partial R}\left(-3H^2R^2\cos\theta\frac{\partial\bar{P}}{\partial R}\right)+\frac{1}{R}\frac{\partial}{\partial\theta}\left(\frac{H^3}{R}\frac{\partial\bar{P}_{yr}}{\partial\theta}\right)+\frac{1}{R}\frac{\partial}{\partial\theta}\left(-3H^2\cos\theta\frac{\partial\bar{P}}{\partial\theta}\right)$$
$$=-K_1\frac{\partial(R\cos\theta)}{\partial\theta}-K_2\frac{\partial}{R\partial R}\left(3H^2R^3\cos\theta\right) \quad (8a)$$

$$\frac{\partial}{R\partial R}\left(RH^3\frac{\partial\bar{P}_{yi}}{\partial R}\right)+\frac{1}{R^2}\frac{\partial}{\partial\theta}\left(H^3\frac{\partial\bar{P}_{yi}}{\partial\theta}\right)=-2K_1\omega_1R\cos\theta \quad (8b)$$

where $R=\frac{r}{r_o}$, $\bar{P}=\frac{p_0}{p_w}$, $H=\frac{h_0}{h_i}$, $\omega_1=\frac{f}{\omega}$, $\bar{P}_{zr}=\frac{p_{zr}h_i}{p_w}$, $\bar{P}_{zi}=\frac{p_{zi}h_i}{p_w}$, $\bar{P}_{xr}=\frac{p_{xr}h_i}{p_w r_o}$, $\bar{P}_{xi}=\frac{p_{xi}h_i}{p_w r_o}$, $\bar{P}_{yr}=\frac{p_{yr}h_i}{p_w r_o}$, $\bar{P}_{yi}=\frac{p_{yi}h_i}{p_w r_o}$, $K_1=\frac{6\mu\omega r_o^2}{p_w h_i^2}$, $K_2=\frac{3\rho\omega^2 r_o^2}{10p_w}$. $\omega$ is the rotating speed. $h_i$ is the film thickness. When $K_2$ equals 0, equations from equation (5) to equation (8b) are the steady and dynamic Reynolds equations without considering centrifugal inertia.

The following boundary conditions are used to solve the equations.

$$\bar{P}=\bar{P}_i=p_i/p_w \text{ at } R=R_i=r_i/r_o\,,$$

$$\bar{P}=\bar{P}_w=1 \text{ at } R=R_o=1\,,$$

$$\bar{P}(\theta,R)=\bar{P}(\theta+2\pi,R)\,,$$

$$\bar{P}_{zr}=\bar{P}_{zi}=\bar{P}_{xr}=\bar{P}_{xi}=\bar{P}_{yr}=\bar{P}_{yi}=0 \text{ at } R=r_i/r_o \text{ and } R=1\,.$$

Equation (5) is a linear equation that governs the steady-state pressure distribution at the equilibrium position. The solution is obtained by a Galerkin finite element method, and then it is used as the input to solve the remaining equations.

The perturbation pressure $\Delta p$ is viewed as a function of perturbation displacement and perturbation velocity of the static ring. Using a first order Taylor series expansion, the perturbation pressure can be expressed as

$$\Delta p=p_{zz}\Delta z+p_{xx}\Delta\gamma_x+p_{yy}\Delta\gamma_y+p_{\dot{z}}\Delta\dot{z}+p_{\dot{x}}\Delta\dot{\gamma}_x+p_{\dot{y}}\Delta\dot{\gamma}_y \quad (9)$$

The axial force and angular moments in response to the perturbation pressure can be expressed in the following form

$$\begin{Bmatrix} dF_z\\ dM_x\\ dM_y\end{Bmatrix}=\iint_A\begin{Bmatrix}1\\ r\sin\theta\\ -r\cos\theta\end{Bmatrix}\Delta p dA=\iint_A\begin{Bmatrix}1\\ r\sin\theta\\ -r\cos\theta\end{Bmatrix}\left(p_{zz}\Delta z+p_{xx}\Delta\gamma_x+p_{yy}\Delta\gamma_y+p_{\dot{z}}\Delta\dot{z}+p_{\dot{x}}\Delta\dot{\gamma}_x+p_{\dot{y}}\Delta\dot{\gamma}_y\right)dA$$
$$=\iint_A\begin{Bmatrix}1\\ r\sin\theta\\ -r\cos\theta\end{Bmatrix}\left[\left(p_{zz}+ifp_{\dot{z}}\right)|\Delta z|e^{ift}+\left(p_{xx}+ifp_{\dot{x}}\right)|\Delta\gamma_x|e^{ift}+\left(p_{yy}+ifp_{\dot{y}}\right)|\Delta\gamma_y|e^{ift}\right]dA \quad (10)$$
$$=\iint_A\begin{bmatrix}p_{zz} & p_{xx} & p_{yy}\\ r\sin\theta p_{zz} & r\sin\theta p_{xx} & r\sin\theta p_{yy}\\ -r\cos\theta p_{zz} & -r\cos\theta p_{xx} & -r\cos\theta p_{yy}\end{bmatrix}dA\begin{Bmatrix}\Delta z\\ \Delta\gamma_x\\ \Delta\gamma_y\end{Bmatrix}+\iint_A\begin{bmatrix}p_{\dot{z}} & p_{\dot{x}} & p_{\dot{y}}\\ r\sin\theta p_{\dot{z}} & r\sin\theta p_{\dot{x}} & r\sin\theta p_{\dot{y}}\\ -r\cos\theta p_{\dot{z}} & -r\cos\theta p_{\dot{x}} & -r\cos\theta p_{\dot{y}}\end{bmatrix}dA\times\begin{Bmatrix}\Delta\dot{z}\\ \Delta\dot{\gamma}_x\\ \Delta\dot{\gamma}_y\end{Bmatrix}$$

Compared with equation (4), the real parts and imaginary parts of perturbation pressures can be expressed as

$$p_{zr}=p_{zz}\,,\ p_{zi}=fp_{\dot{z}}\,,\ p_{xr}=p_{xx}\,,\ p_{xi}=fp_{\dot{x}}\,,$$

$$p_{yr}=p_{yy}\,,\ p_{yi}=fp_{\dot{y}}\,.$$

IOP Conf. Series: Materials Science and Engineering **378** (2018) 012017 doi:10.1088/1757-899X/378/1/012017

Introducing the following form

$$\begin{Bmatrix} \mathrm{d}F_z \\ \mathrm{d}M_x \\ \mathrm{d}M_y \end{Bmatrix} = -\begin{bmatrix} k_{zz} & k_{zx} & k_{zy} \\ k_{xz} & k_{xx} & k_{xy} \\ k_{yz} & k_{yx} & k_{yy} \end{bmatrix} \times \begin{Bmatrix} \Delta z \\ \Delta\gamma_x \\ \Delta\gamma_y \end{Bmatrix} - \begin{bmatrix} c_{zz} & c_{zx} & c_{zy} \\ c_{xz} & c_{xx} & c_{xy} \\ c_{yz} & c_{yx} & c_{yy} \end{bmatrix} \times \begin{Bmatrix} \Delta\dot{z} \\ \Delta\dot{\gamma}_x \\ \Delta\dot{\gamma}_y \end{Bmatrix} \tag{11}$$

Comparing equation (10) with equation (11), the dimensionless stiffness and damping coefficients are easily identified

$$\begin{bmatrix} K_{zz} & K_{zx} & K_{zy} \\ K_{xz} & K_{xx} & K_{xy} \\ K_{yz} & K_{yx} & K_{yy} \end{bmatrix} = -\iint_A \begin{bmatrix} \overline{P}_{zr} & \overline{P}_{xr} & \overline{P}_{yr} \\ R\sin\theta\overline{P}_{zr} & R\sin\theta\overline{P}_{xr} & R\sin\theta\overline{P}_{yr} \\ -R\cos\theta\overline{P}_{zr} & -R\cos\theta\overline{P}_{xr} & -R\cos\theta\overline{P}_{yr} \end{bmatrix} \mathrm{d}A \tag{12}$$

$$\begin{bmatrix} C_{zz} & C_{zx} & C_{zy} \\ C_{xz} & C_{xx} & C_{xy} \\ C_{yz} & C_{yx} & C_{yy} \end{bmatrix} = -\frac{1}{\omega_1}\iint_A \begin{bmatrix} \overline{P}_{zi} & \overline{P}_{xi} & \overline{P}_{yi} \\ R\sin\theta\overline{P}_{zi} & R\sin\theta\overline{P}_{xi} & R\sin\theta\overline{P}_{yi} \\ -R\cos\theta\overline{P}_{zi} & -R\cos\theta\overline{P}_{xi} & -R\cos\theta\overline{P}_{yi} \end{bmatrix} \mathrm{d}A \tag{13}$$

where $K_{zz} = \frac{k_{zz}h_i}{p_w r_o^2}$, $K_{zx} = \frac{k_{zx}h_i}{p_w r_o^3}$, $K_{zy} = \frac{k_{zy}h_i}{p_w r_o^3}$, $K_{xz} = \frac{k_{xz}h_i}{p_w r_o^3}$, $K_{xx} = \frac{k_{xx}h_i}{p_w r_o^4}$, $K_{xy} = \frac{k_{xy}h_i}{p_w r_o^4}$, $K_{yz} = \frac{k_{yz}h_i}{p_w r_o^3}$, $K_{yx} = \frac{k_{yx}h_i}{p_w r_o^4}$, $K_{yy} = \frac{k_{yy}h_i}{p_w r_o^4}$, $C_{zz} = \frac{c_{zz}h_i\omega}{p_w r_o^2}$, $C_{zx} = \frac{c_{zx}h_i\omega}{p_w r_o^3}$, $C_{zy} = \frac{c_{zy}h_i\omega}{p_w r_o^3}$, $C_{xz} = \frac{c_{xz}h_i\omega}{p_w r_o^3}$, $C_{xx} = \frac{c_{xx}h_i\omega}{p_w r_o^4}$, $C_{xy} = \frac{c_{xy}h_i\omega}{p_w r_o^4}$, $C_{yz} = \frac{c_{yz}h_i\omega}{p_w r_o^3}$, $C_{yx} = \frac{c_{yx}h_i\omega}{p_w r_o^4}$, $C_{yy} = \frac{c_{yy}h_i\omega}{p_w r_o^4}$.

## 3. Results and discussion

The calculation results show that angular coupling stiffness coefficients $K_{zx}$, $K_{zy}$, $K_{xz}$, $K_{yz}$ and angular coupling damping coefficients $C_{zx}$, $C_{zy}$, $C_{xz}$, $C_{yz}$ are all equal to zero. Angular stiffness coefficients $K_{xx}=K_{yy}$, $K_{xy}=-K_{yx}$. Angular damping coefficients $C_{xx}=C_{yy}$, $C_{xy}=C_{yx}=0$.

### *3.1. Effects of centrifugal inertia under different rotating speed*

Figure 3 shows the effects of centrifugal inertia on dynamic stiffness coefficients under different rotating speed. It can be seen that the dynamic stiffness coefficients increase linearly with the increase of rotating speed. The effect of rotating speed on axial stiffness coefficient $K_{zz}$ is greater than that on angular stiffness coefficient $K_{xx}$ ($K_{yy}$). However, the effect of rotating speed on angular coupling stiffness coefficient $K_{xy}$ ($-K_{yx}$) is the smallest. At the same time, it can be found that the dynamic stiffness coefficients with centrifugal inertia are basically the same as those without centrifugal inertia. The relative errors increase linearly with the increase of rotating speed. When the rotating speed is 12000 $\mathrm{r\cdot min^{-1}}$, the relative error is respectively 0.68%, 0.81% and 0.19%, which can be neglected.

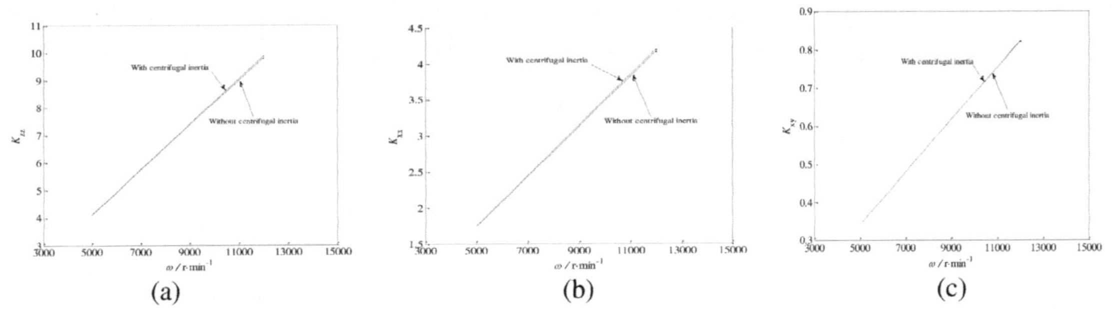


(a) (b) (c)

**Figure 3.** Effects of centrifugal inertia on dynamic stiffness coefficients under different rotating speed.

Figure 4 describes the effects of centrifugal inertia on dynamic damping coefficients under different rotating speed. The figures show that the dynamic damping coefficients increase with the increase of rotating speed, which is basically linear. The effect of rotating speed on axial damping coefficient $C_{zz}$ is greater than that on angular damping coefficient $C_{xx}$ ($C_{yy}$). Compared the dynamic damping coefficients with centrifugal inertia and those without centrifugal inertia under the same rotating speed, it is known that the two are exactly the same, so centrifugal inertia has no effect on dynamic damping coefficients when rotating speed changes.

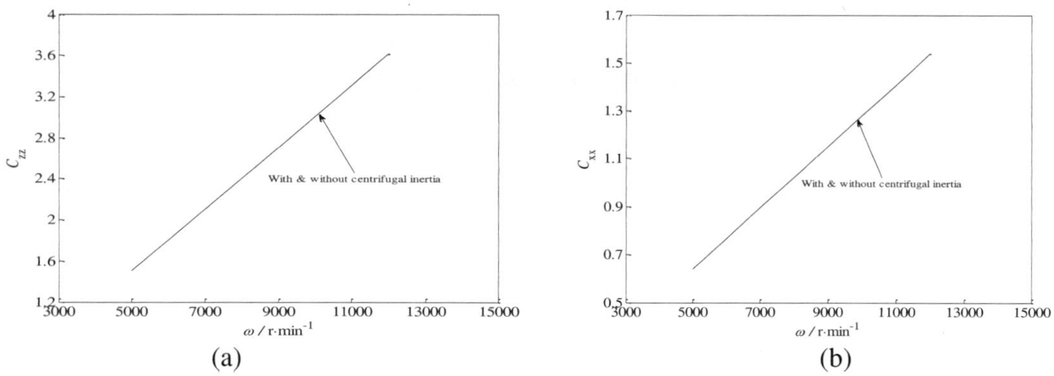


**Figure 4.** Effects of centrifugal inertia on dynamic damping coefficients under different rotating speed.

*3.2. Effects of centrifugal inertia under different film thickness*

Figure 5 shows the effects of centrifugal inertia on dynamic stiffness coefficients under different film thickness. From the figures, we know that axial stiffness coefficient $K_{zz}$ and angular stiffness coefficient $K_{xx}$ ($K_{yy}$) decrease rapidly with the increase of film thickness firstly, and then the trend of decline is slow and tends to zero. Besides, the change of $K_{zz}$ is more significant than that of $K_{xx}$ ($K_{yy}$). The angular coupling stiffness coefficient $K_{xy}$ ($-K_{yx}$) decreases gradually and tends to zero as the film thickness increases. Compared the dynamic stiffness coefficients with centrifugal inertia and those without centrifugal inertia, we can know that the relative errors increase as the film thickness increases, and the relative error is respectively 3.26%、5.08% and 0.26% when the film thickness is about $15\times10^{-6}$ m.

Figure 6 gives the effects of centrifugal inertia on dynamic damping coefficients under different film thickness. The figures show that axial damping coefficient $C_{zz}$ and angular damping coefficient $C_{xx}$ ($C_{yy}$) are decreased rapidly with the increase of film thickness and the change of $C_{zz}$ is more obvious. With a further increase of film thickness, axial damping coefficient and angular damping coefficient decrease slowly and tend to zero. The dynamic damping coefficients with centrifugal inertia are exactly the same as those without centrifugal inertia under the same film thickness. Therefore, centrifugal inertia has no effect on dynamic damping coefficients when film thickness changes.

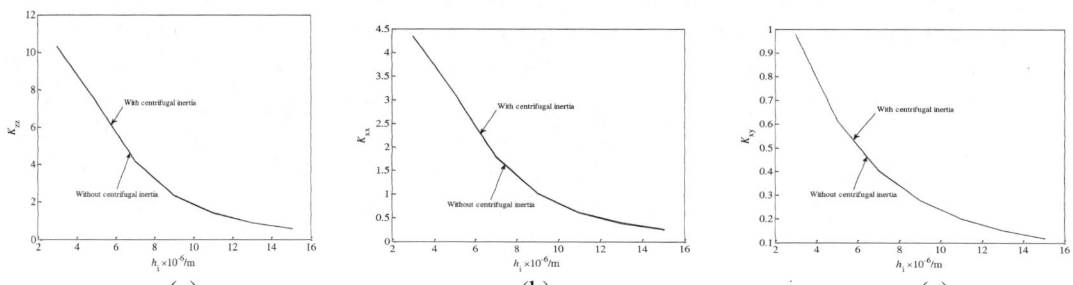


**Figure 5.** Effects of centrifugal inertia on dynamic stiffness coefficients under different film thickness.

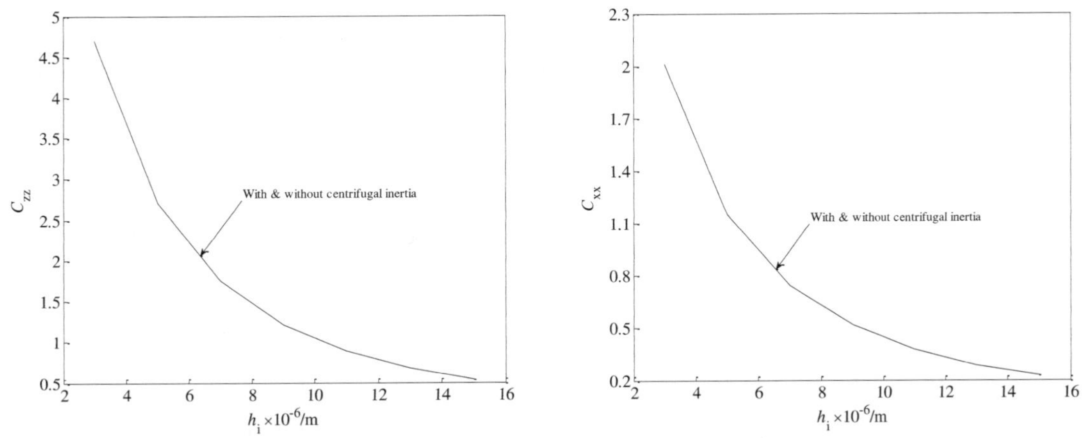


(a) (b)

**Figure 6.** Effects of centrifugal inertia on dynamic damping coefficients under different film thickness.

## 4. Conclusions

In this paper, the stiffness and damping coefficients are obtained by solving the dynamic Reynolds equation using the perturbation technique. Results show that the effects of centrifugal inertia on axial stiffness coefficient and angular coupling stiffness coefficient can be neglected under different rotating speed and different film thickness. When the values of rotating speed and film thickness are large, the effect of centrifugal inertia on angle stiffness coefficient cannot be ignored. Centrifugal inertia has no effects on dynamic damping coefficients with the variation of rotating speed and film thickness.

It should be noted that the fluid pressure in this work is greater than cavitation pressure, and a cavitation phenomenon doesn't takes place. However, when the film thickness is small or the fluid is oil, the fluid pressure may be less than cavitation pressure, which should be studied in the future.

## 5. Acknowledgment

The authors gratefully acknowledge the financial support by the National Natural Science Foundation of China (51375497), and the Graduate Innovation Project of China University of Petroleum (East China) (YCXJ2016040).

## 6. References

[1] Salant R F and Homiller S J 1993 Stiffness and leakage in spiral groove upstream pumping mechanical seals. *Tribology Transactions*, **36** 55-60.

[2] Kakoty S K and Majumdar B C 1999 Effect of fluid inertia on stability of flexibly supported oil journal bearings: linear perturbation analysis. *Tribology International*, **32** 217-28.

[3] Nassab G S A 2005 Inertia effect on the thermohydrodynamic characteristics of journal bearings. *Proceedings of the Institution of Mechanical Engineers, Part J: Journal of Engineering Tribology*, **219** 459-67.

[4] Pinkus O and Lund J W 1981 Centrifugal effects in thrust bearings and seals under laminar conditions. *ASME Journal of Lubrication Technology*, **103** 126-36.

[5] Gupta R S and Sharma L G 1989 Centrifugal inertia effects in misaligned radial face seals. *Wear*, **129** 319-32.

[6] Zhao Y M, Hu J B and Wei C 2014 Dynamic analysis of spiral-groove rotary seal ring for wet clutches. *Journal of Tribology*, **136** 0317101-10.

[7] Zhao Y M, Yuan S H, Hu J B and Wei C 2015 Nonlinear dynamic analysis of rotary seal ring considering creep rotation. *Tribology International*, **82** 101-09.

[8] Ruan B 2002 A semi-analytical solution to the dynamic tracking of non-contacting gas face seals. *Journal of Tribology*, **124** 196-202.
[9] HimYu T and Sadeghi F 2001 Groove effects on thrust washer lubrication. *Journal of Tribology*, **123** 295-304.

IOP Conf. Series: Materials Science and Engineering **378** (2018) 012018 doi:10.1088/1757-899X/378/1/012018

# Research on Surface Topography Characteristics of Laser Micro-hole Drilling

**H J Ding, X H Li and D Wang**

College of Mechanical Engineering Jiamusi University, Jiamusi, Heilongjiang, China

E-mail: dhj1000@126.com

**Abstract.** The surface of laser micro processing has great differences compared with the traditional machining surface. To research surface topography is of great significance for laser processing surface property. Combined with the laser mechanism, the surface characteristics of micro blind-hole drilled with different pulse energy and number were studied in the materials of spring steel 65Mn, stainless steel SUS301 and ceramic $Al_2O_3$. The results indicate that large pulse energy makes the hole much irregular and multi-pulse may cause the surface much rough. A large amount of molten material gathers around the hole for 65Mn and SUS301 while not for $Al_2O_3$ because of different processing mechanism. Therefore, process parameters and conditions should be explored for better surface topography of laser drilling.

## 1. Introduction

Laser micro processing is a kind of advanced and nontraditional machining technology. Its surface is special which is between the isotropic and the anisotropic. The surface topography can reflect the producing mechanism objectively and also influence the functional feature of engineering surface, such as the friction and lubrication property [1-2].

Laser drilling is a complex photothermal conversion process. The surface topography is affected directly by the laser absorption and reflection of material and heat transfer [3-4]. The research on the surface quality of laser micro machining is neither systematic nor comprehensive, and not evaluated at present, which seriously restricts the development of laser processing technology [5]. So it is important and meaningful to research the laser micro machining surface topography.

## 2. Experiment scheme

The micro blind-holes were drilled by laser in the materials of stainless steel SUS301, spring steel plate 65Mn and ceramic $Al_2O_3$ with thickness of 0.3mm, voltage of 750v, 800v, 850v and 900v and different pulse number respectively. The influence of laser pulse energy, pulse numbers and different materials on surface topography was studied by using scanning electron microscope (SEM) [6].

## 3. Experiment results and analysis

### *3.1. Influence of pulse energy on micro-hole surface topography of laser drilling*

Figure 1 shows the surface topography SEM images of micro blind-holes with the single pulse laser under different voltage. From Figure 1(a), when the pulse energy is small, the hole trace is not obvious. Increasing the energy, the laser hole becomes clear and similar to the circle, as shown in Figure 1(b). The instantaneous high temperature produced by laser drilling only can melt and liquefy but less gasify, so the main style of material is molten metal. The topography of hole is like the ripple of water when the molten metal is impacted and spread around by the laser pulse. When the pulse energy continues to increase, as shown in Figure 1 (c) and (d), part of the material is sublimated and gasified. The gaseous material mixed with the molten metal is ejected in a high speed from the hole. The volume of

IOP Conf. Series: Materials Science and Engineering **378** (2018) 012018 doi:10.1088/1757-899X/378/1/012018

material increases rapidly in a very short time and produces a local explosion effect. The explosion force causes the molten metal to flow around and piles up and then forms a protruding outer edge around the hole. The explosion of gasification metal is random and the molten material is ejected from each direction which makes holes irregular circle.

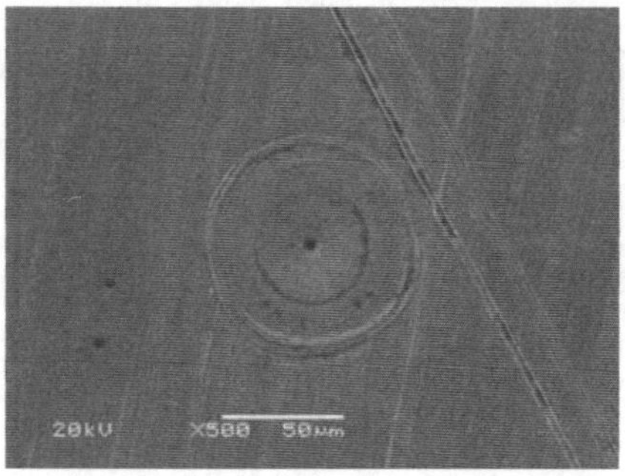


(a) 52mJ

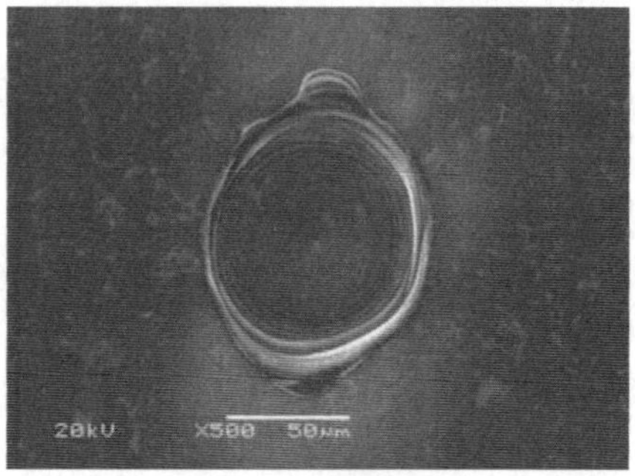


(b) 56mJ

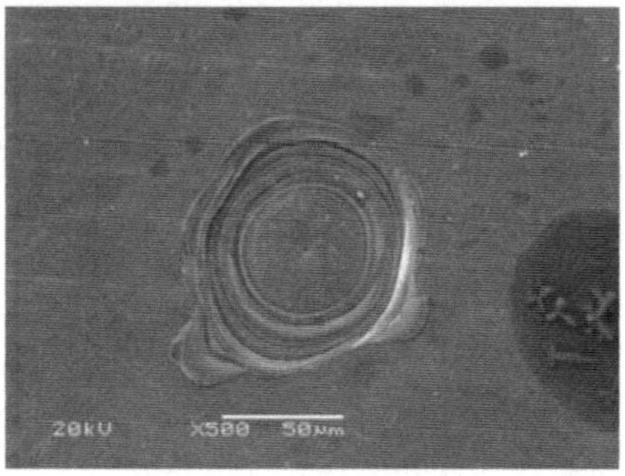


(c) 61mJ

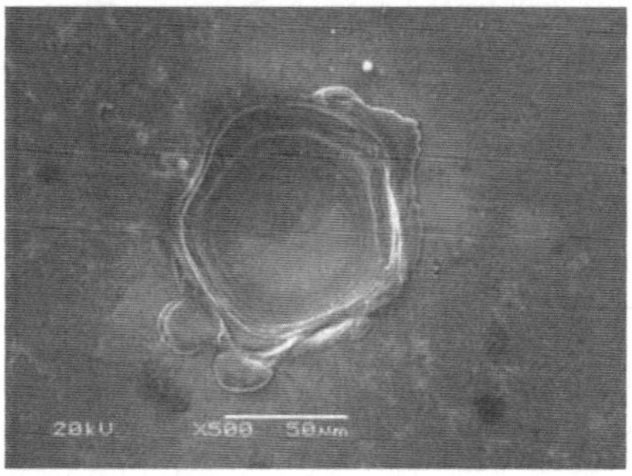


(d) 65mJ

**Figure 1.** Surface topography of micro blind-hole with the single pulse

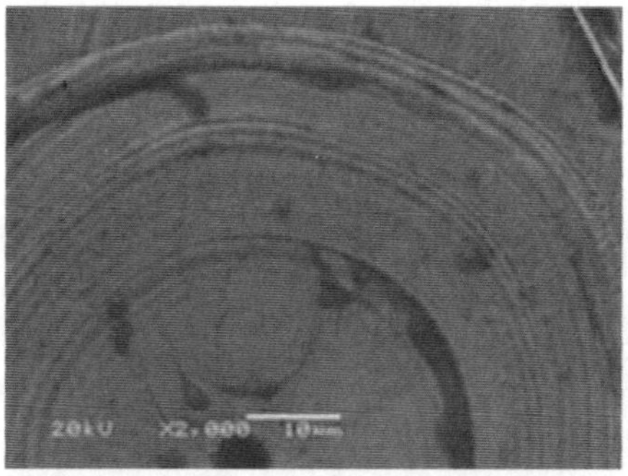


**Figure 2.** SUS301 800v single pulse

The laser beam is a Gauss beam, and the energy maximum is in the center. The center material of hole is all vaporized and removed and makes the topography smooth relatively. The molten metal is cooled rapidly and accumulated in the hole wall and forms concentric circles after laser pulse, as shown in Figure 2.

*3.2. Influence of pulse number on micro-hole surface topography of laser drilling*

With the increase of pulse number, laser energy is larger. Figure 3 shows that the concentric circles are formed in the wall of holes. The material is removed out more and more with metal gasification. Some big slags are exploded around the hole by the beginning pulse and the molten metal is extruded by the subsequent. The flow of liquid metal is blocked by the slag and formed the accumulation tumor, as shown in Figure 3 (c) and (d).

IOP Conf. Series: Materials Science and Engineering **378** (2018) 012018 doi:10.1088/1757-899X/378/1/012018

The molten pool has appeared on the surface of the material with the single pulse and it becomes deeper from outer to inner layer by layer. The peripheral hole is concentric circles formed by a number of molten pools. Obviously, under such processing condition, it is not enough to make the material melted completely, and the surface of hole is extremely rough. The hole becomes relatively uniform by using two or three pulses, and the bottom has become smooth. But it is not complete to melt yet. When the number of pulse is four, the bottom of the hole becomes much uniform and arc surface is inerratic, too. But the arc surface of the hole becomes worse when drilling with 5 pulses though the bottom is still smooth. Thus, it is much appropriate by using 4 pulses to make blind-hole inerratic and uniform.

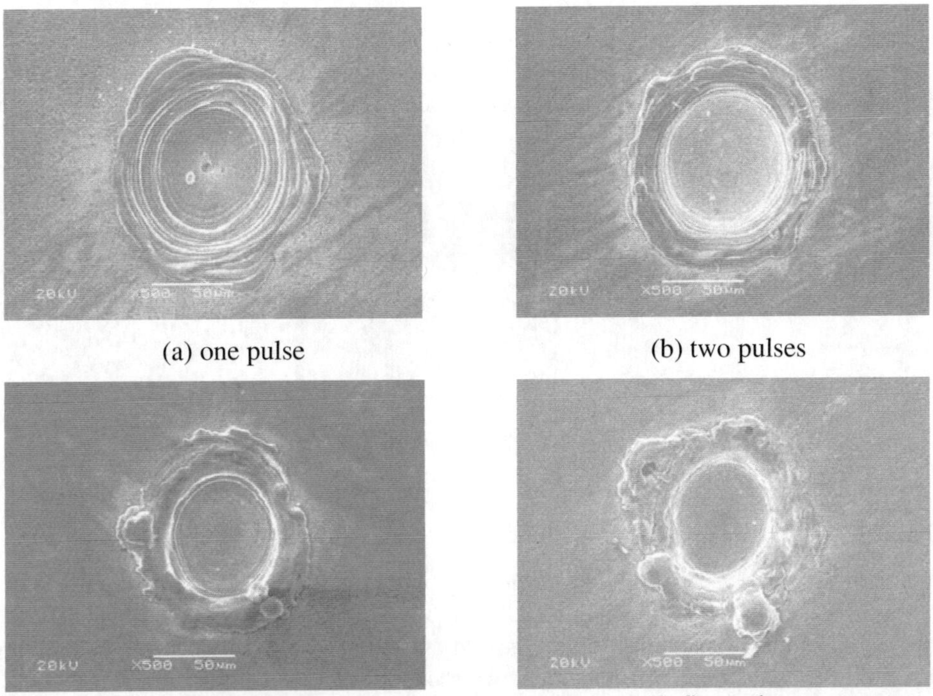

(a) one pulse (b) two pulses

(c) four pulses (d) five pulses

**Figure 3.** Micro-hole topography of different pulse number with the same voltage.

Figure 4 shows the hole diameter is becoming smaller and smaller and then the hole may be blocked when drilling with multi-pulse. This is because the heat generated by the laser pulse makes the material melt and gasify. The depth increases with the material flying outward from the hole. The focus of the laser doesn't gather at the bottom of hole, but in the middle. The energy which can reach the bottom reduces and makes material removal reduce too. A large amount of material gathers at the entrance of hole, and blocks the hole after cooling. So the ideal topography of hole will not be obtained if continue to drill with this processing condition.

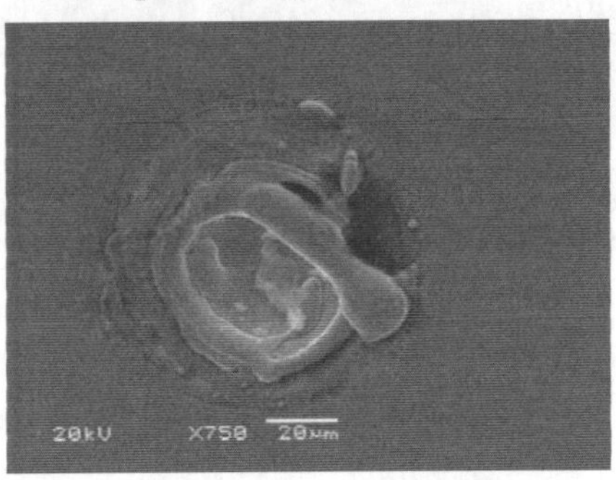

**Figure 4.** SUS301 750v 6 pulses.

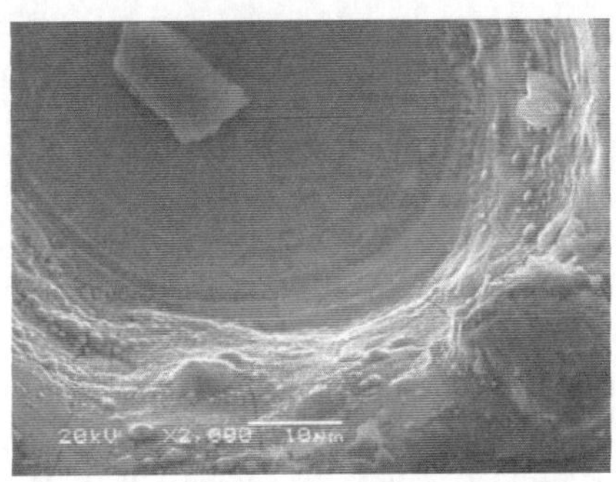

**Figure 5.** SUS301 800v 19 pulses.

Some large pieces of slag are discharged when drilling with high pulse energy and multi-pulse. Some slag falls back into the hole and bonds with the bottom after cooling. When slag is produced too much, some rough peaks and valleys generate in the wall and bottom of hole which could affect the surface topography seriously, as shown in Figure 5. An amount of gas is generated and metal bubble is caused around the hole wall. When the energy is sufficient, many metal bubbles which can not escape in time are adhered to the inner wall of the hole. The temperature difference causes tension stress on the surface of material. When the tension stress exceeds the strength of extension, many small cracks are produced in the wall of the hole which can affect the service life of the workpiece.

*3.3 Surface topography of laser micro-hole drilling with different material*
Figure 6 shows the laser drilling SEM images of different material with the same pulse energy and number. In order to avoid most of the energy is reflected and lost through the workpiece, it was blackened before drilling in ceramic $Al_2O_3$, and then the blackened material was washed with ultrasonic waves after processing.

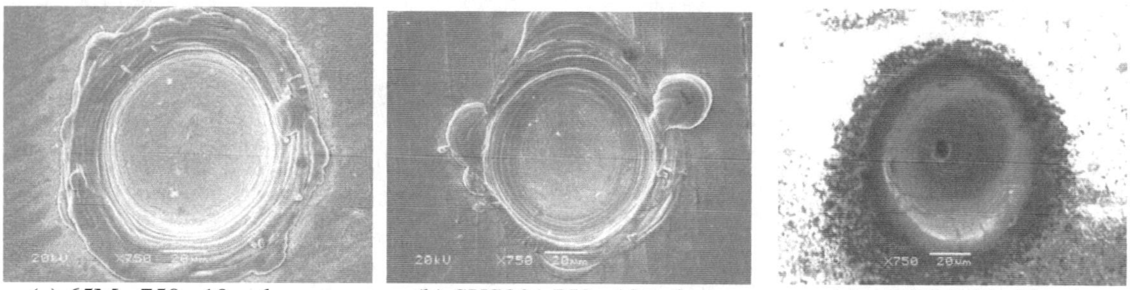

(a) 65Mn 750v 10 pulses (b) SUS301 750v 10 pulses (c) Al2O3 750v 10 pulses

**Figure 6.** SEM images of different material with the same pulse energy and numbers.

It can be seen from Figure 6 that there are a large amount of molten material which gathers around the hole but the bottom arc is flat for 65Mn and SUS301. For $Al_2O_3$ ceramic, the molten material around hole does not gather into a boss and the hole is thin and sharp. That is because the laser processing mechanism of ceramic is different from metal material. 65Mn and SUS301 belong to plastic metal material, and removal mainly depends on the thermal effect. While $Al_2O_3$ ceramic is brittle and material remove mainly depends on ablative photo decomposition (APD).

## 4. Summary

The effect of pulse energy, pulse number and different material on the surface topography of laser micro-hole was investigated using SEM. The experimental results show that the irregular circle was obtained because of the explosive force of molten metal when the pulse energy became greater. A certain number of laser pulses can get better quality of hole. But too many pulses cause uneven peaks and valleys in the wall and bottom of hole which affect the surface topography, and even make the hole blocked by molten slag. The processing mechanism is different between ceramic and metal material, and their surface topography has a great difference. These results have laid a foundation for further study of three-dimensional roughness parameter evaluation of laser micro machining surface.

## 5. Acknowledgments

This work was financially supported by National Natural Science Foundation of China (51375208), research fund project of Jiamusi University (Lz2014-002) and precision manufacturing technology research team project of Jiamusi University (Cxtdpy-2016-04).

## 6. References

[1] Romoli L, Rashed C.A.A, Lovicu G and Dini G 2014 *J. Manu. Technol.* **63** 229-32
[2] Arrizubieta I, Martínez S and Ukar E 2013 *J. Mach. Tool. and Manu.* **75** 55-62
[3] Mishra S and Yadava V 2013 *J. Eng. Manu.* **227** 494-507

[4] Yan Y, Ji L, Bao Y and Jiang Y 2012 *J. Mater. Proce. Technol.* **212** 1257-70
[5] Romoli L, Rashed C.A.A and Fiaschi M 2014 *J. Opti.& Las. Technol.* **56** 35-42
[6] Nadolny K and Kaplonek W 2012 *J. Mecha. Eng.and Rob. Resear.* **1** 14-29

IOP Conf. Series: Materials Science and Engineering **378** (2018) 012019 doi:10.1088/1757-899X/378/1/012019

# Research on AC Power Conditioning Control Technology for Ultraprecision Machining of Steel Materials

**En-Chih Chang**

Department of Electrical Engineering, I-Shou University, Taiwan, R.O.C.
E-mail: enchihchang@isu.edu.tw

**Abstract.** This paper presents a nonlinear control technology controlled AC power conditioning with application to ultraprecision machining of steel materials. The presented technology associates the advantages of finite-time tracking control (FTTC) and cuckoo search algorithm (CSA). The FTTC allows insensitivity to system uncertainties as well as system states finite-time convergence. It is a remarkable fact that the chatter will occur in face of highly dynamic loads. The chatter causes high output-voltage distortion in AC power conditioning, and the ultraprecision machining of steel materials may be instability and unreliability. The CSA is thus used to attenuate the chatter so that the AC power conditioning can provide robust performance for ultraprecision machining of steel materials. Because the proposed control technology is easier to implement than prior technologies and achieves high tracking precision and low calculational-complexity algorithm, experiments display low total harmonic distortion and fast transience in the output voltage, and this paper will be helpful to researchers of related ultraprecision machining of steel materials.

## 1. Introduction

An AC power conditioning has been used as a key element for ultraprecision machining of steel materials, like rolling, forging, and drawing [1]. The high performance AC power conditioning should contain 1) the output voltage waveform with low harmonic distortion for linear or nonlinear cyclic loads. 2) the fast transient response when the abrupt load changes occurs. 3) the steady-state error should be restricted as small as possible. To gain requisitions, proportional-integral-derivative controllers perform well in the presence of constant load circumstances; however, the good performance cannot be satisfied when the plant is subject to fluctuating load disturbances [2], [3]. Nonlinear control strategies are frequently adopted, such as direct repetitive control [4], optimal state feedback [5], and adaptive hysteresis control [6]. But, they have complex algorithms, and implemented difficulty. Variable structure control (VSC) is a robust method for controlling nonlinear system where occurs system uncertainties [7], [8]; a number of VSC developed for AC power conditioning have been done, but linear sliding functions are employed, and infinite-time tracking problem exists [9], [10]. To enhance the convergence speed, the FTTC with nonlinear sliding function is used to make tracking error converge to zero in finite time [11], [12]. However, once a highly dynamic load occurs, the chatter appears in AC power conditioning output with high harmonic distortion, thus deteriorating the stability and reliability of ultraprecision machining of steel materials. The cuckoo search algorithm (CSA) is a swarm intelligence method which inspired by the unique breeding behavior of cuckoos, and has been presented for solving global optimization. Several works have improved the chatter by the use of the optimal approaches [13]; however, these have complex calculation and slow convergence rate. Therefore, the CSA is used to attenuate the chatter while the

FTTC provides system states finite-time convergence. By the combination of FTTC and CSA, the tracking error between desired and actual output will be minimized [14], [15], and the AC power conditioning of ultraprecision machining of steel materials yields low total harmonic distortion, fast dynamic response, chatter attenuation, and steady-state errors reduction. Experiments are given to verify the applicability and efficacy of using the proposed technology.

## 2. Presented Control Technology for Ultraprecision Machining of Steel Materials

Figure 1 shows the system block diagram of AC power conditioning in ultraprecision machining of steel materials. Let $v_o$ be the output voltage, $v_r$ be the desired AC waveform, $v_e = v_o - v_r$ be the voltage error, and the load is represented by $R$. Assume that the switching frequency is high enough, the AC power conditioning can be regarded as a constant gain, $K_{PWM}$. Define $e_1 = v_e$ and $\dot{e}_1 = e_2$, the tracking error can be expressed as the following matrix

$$\begin{cases} \dot{e}_1 = e_2 \\ \dot{e}_2 = -a_1 e_1 - a_2 e_2 + bv - N \end{cases} \tag{1}$$

where $a_1$ is $1/LC$, $a_2$ denotes $1/RC$, $b$ stands for $K_{PWM}/LC$, and $N = v_r/LC + \dot{v}_r/RC + \ddot{v}_r$ symbols system uncertainty.

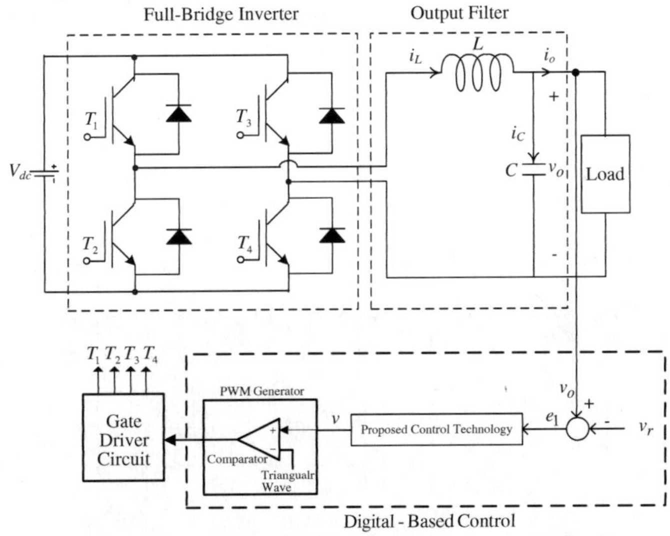


**Figure 1.** AC power conditioning in ultraprecision machining of steel materials.

A finite-time tracking control with nonlinear sliding function is defined as

$$s = e_1 + (1/\rho)\cdot e_2^{p/q} \tag{2}$$

where $p > q$, $p$ and $q$ are positive odd numbers ($1 < p/q < 2$), and a sliding-mode reaching item $\dot{s} = -ks - \varepsilon sign(s)$ is used. Therefore, the control law $v$ can be written as

$$v = \underbrace{b^{-1}[a_1 e_1 + a_2 e_2 - (\rho q/p)\cdot e_2^{2-p/q}]}_{u_{eq}} \underbrace{- b^{-1}[ks + (L_g + \varepsilon)sign(s)]}_{u_{ft}}, \; k > 0,\ \varepsilon > 0 \tag{3}$$

where $u_{eq}$ indicates the equivalent control which determines the system dynamics, and $u_{ft}$ signifies the sliding control which prevents the system uncertainties. Thus, the system will be driven to $s = 0$ and converged in finite time; however, if a highly dynamic load is applied, the chatter still occurs, and the system (1) cannot obtain precise tracking performance. To attenuate the chatter, the control gains in (3) can be optimally tuned by the use of the CSA, and the concept of CSA is briefly described as follows:

IOP Conf. Series: Materials Science and Engineering **378** (2018) 012019 doi:10.1088/1757-899X/378/1/012019

(i) Each cuckoo lays an egg at one time, and gets rid of its egg in a randomly select nest.
(ii) The best nests have high-quality eggs that keep to the next generation.
(iii) Let the number of accessible host nests be fixed, and we suppose that a host bird recognizes the cuckoo egg with the probability of $p_a \in [0,1]$ so that the host bird can either discard them or abandon to create a new nest in a new location.

Once a new solution $x_i(t+1)$ is produced for a cuckoo $i$, a Levy flight can be constructed as

$$x_i(t+1) = x_i(t) + \gamma \oplus \text{Levy}(\lambda) \quad (4)$$

where $\gamma$ is the step size, and the product $\oplus$ represents entry wise multiplications. Also, there is a random walk with random step size in the Levy flight that follows a Levy's distribution below.

$$\text{Levy} \sim u = t^{-\lambda},\ 1 < \lambda \le 3 \quad (5)$$

## 3. Experiments

The proposed system parameters are listed as follows: $V_{dc}$=200 V; $v_o$=110 $V_{rms}$, $f$=60 Hz; $L$=0.1 mH; $C$=3 μF; Switching Frequency $f_s$ =15 kHz; $R_{rated}$ =12 ohm. To test the transient behavior for ultraprecision machining of steel materials, Fig. 2 and Fig. 3 show the output voltage and the load current obtained using the proposed control technology and the classic VSC under step change in load (from no load to full load) at a 90 degree firing angle, respectively. A detailed examination of the waveforms shows that Fig. 2 has tiny voltage sag and fast retrieval of the steady-state response. But, the classic variable structure controlled system, shown in Fig. 3 yields unsatisfactory transient response and delayed recovery time. Fig. 4 and 5 depict the tracking errors in the proposed control technology and classic VSC, respectively. It is worth noting that the tracking error of the proposed controlled system not only shows system states finite-time convergence but also approximately exists vibration-free than that of the classic variable structure controlled system. Therefore, it is well confirmed that the proposed control technology always produces higher tracking exactness, lower harmonic distortion, and faster convergence speed, as comparing the classic VSC.

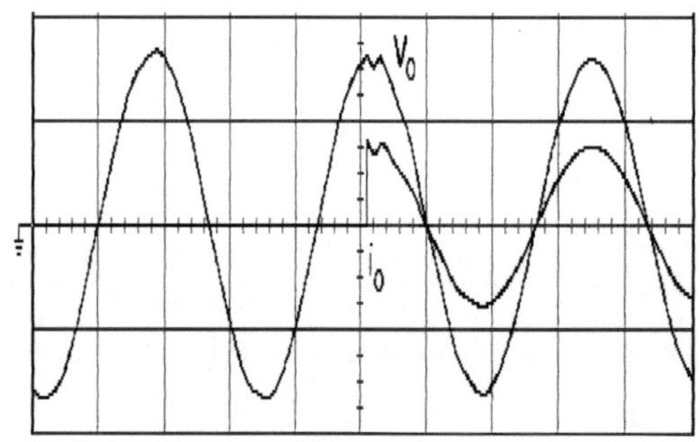


**Figure 2.** Proposed control technology under step change in load (100V/div; 20A/div; 5ms/div).

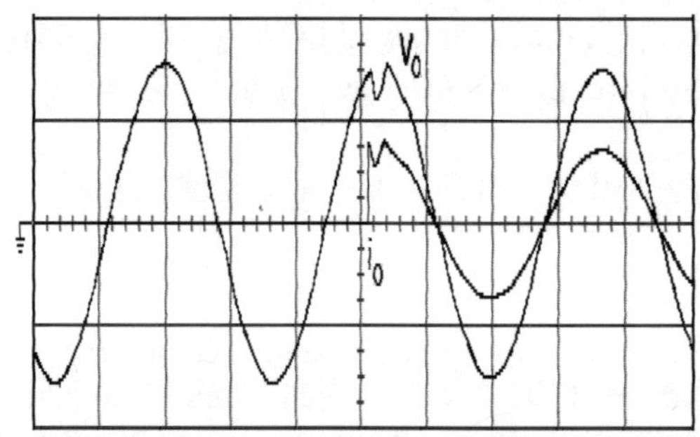


**Figure 3.** Classic VSC under step change in load (100V/div; 20A/div; 5ms/div).

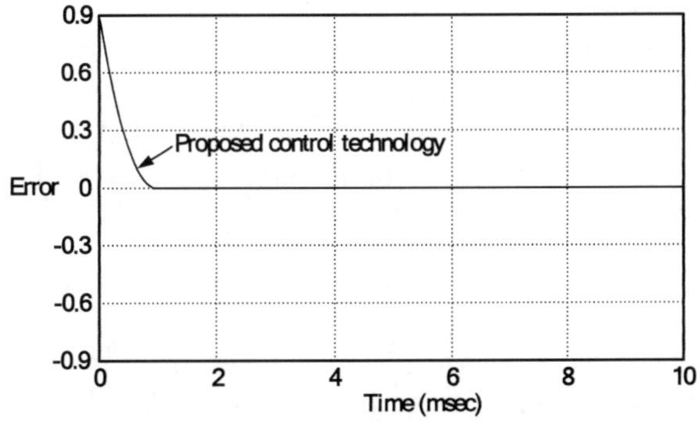


**Figure 4.** Tracking errors in the proposed control technology.

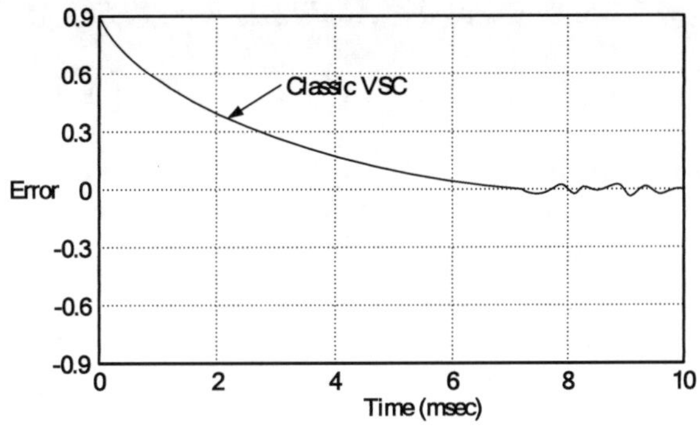


**Figure 5.** Tracking errors in the classic VSC.

## 4. Conclusions

This paper has proposed FTTC with CSA for designing AC power conditioning to improve ultraprecision machining of steel materials. This proposed control technology not only has the robustness of classic VSC but also provides finite time convergence of the system state, and attenuates the chatter. According to theory and experimental results, the validity of the proposed control technology is successfully demonstrated and thus more appropriate for use in ultraprecision machining of steel materials. Further, in future research, the proposed control technology can be used to more complex three phase AC power conditioning for machining applications.

## 5. Acknowledgment

This work was supported by the Ministry of Science and Technology of Taiwan, R.O.C., under contract number MOST106-2221-E-214-031.

## 6. References

[1] Gupta K 2017 *Advanced Manufacturing Technologies: Modern Machining, Advanced Joining, Sustainable Manufacturing* (New York: Springer)

[2] Shin H B and Park J G 2012 Anti-Windup PID Controller with Integral State Predictor for Variable-Speed Motor Drives *IEEE Trans. Industrial Electronics* vol. 59, no. 3 pp 1509-1516

[3] Rasoanarivo I and Sargos F 2013 Multi-Objective Analysis for Designing and Controlling Micro-grids under Multi-Control with PID, MHCC and FOPID Controllers *Proc. IEEE Int. Conf. on Industry Applications Society Annual Meeting* pp 1-8

[4] Zhang K Kang Y Xiong J and Chen J 1995 Direct Repetitive Control of SPWM Inverter for UPS Purpose *IEEE Trans. Power Electronics* vol. 18, no. 3 pp 271-280

[5] Jung S L Ho L H Yeh H C and Tzou YY 1994 DSP-Based Digital Control of a PWM Inverter for Sine Wave Tracking by Optimal State Feedback Technique *Proc. IEEE Int. Conf. on Power Electronics Specialists Conference* pp 546-551

[6] Kawamura A and Hoft R G 1984 Instantaneous Feedback Controlled PWM Inverter with Adaptive Hysteresis *IEEE Trans. Industry Applications* vol. 20, no. 4 pp 769-775

[7] Liu J K and Wang X H 2012 *Advanced Sliding Mode Control for Mechanical Systems Design, Analysis and MATLAB Simulation* (Berlin: Springer-Verlag)

[8] Wu L G Shi P and Su X J 2014 *Sliding Mode Control of Uncertain Parameter-Switching Hybrid Systems* (New York: Wiley)

[9] Banerjee B and Weaver W W 2014 Generalized Geometric Control Manifolds of Power Converters in a DC Microgrid *IEEE Trans. Energy Conversion* vol. 29, no. 4 pp 904-912

[10] Tan S C Lai Y M and Tse C K 2012 *Sliding Mode Control of Switching Power Converters: Techniques and Implementation* (Boca Raton: CRC Press)

[11] Azar A T and Zhu Q M 2015 *Advances and Applications in Sliding Mode Control Systems* (New York: Springer)

[12] Shtessel Y Edwards C Fridman L and Levant A 2014 *Sliding Mode Control and Observation* (New York: Springer)

[13] Aschepkov L T Dolgy D V Kim T and Agarwal R P 2016 *Optimal Control* (New York: Springer)

[14] Yang X S 2014 *Nature-Inspired Optimization Algorithms* (Boston: Elsevier)

[15] Mishra C Singh S P and Rokadia J 2015 Optimal Power Flow in the Presence of Wind Power Using Modified Cuckoo Search, *IET Proceedings-Generation, Transmission & Distribution* vol. 9, no. 7 pp 615-626

**Institute of Physics**
Dirac House, Temple Back
Bristol BS1 6BE UK

ISSN: 1757-8981
ISBN 978-1-5108-6585-3

9 781510 865853

CURRAN ASSOCIATES INC.
proceedings.com

# International Conference on Defects in Insulating Materials (ICDIM 2016)

IOP Conference Series: Materials Science and Engineering
Volume 169

Lyon, France
10 – 15 July 2016

IOP
Institute of Physics